Advances, Applications, and Future Perspectives of Aerogels

Abel Inobeme
Edo University, Uzairue, Nigeria

John Tsado Mathew
Ibrahimm Badamasi Babangida Lapai, Nigeria

Charles Oluwaseun Adetunji
Edo University, Auchi, Nigeria

IGI Global
Scientific Publishing
Publishing Tomorrow's Research Today

Vice President of Editorial Melissa Wagner
Managing Editor of Acquisitions Mikaela Felty
Managing Editor of Book Development Jocelynn Hessler
Production Manager Mike Brehm
Cover Design Phillip Shickler

Published in the United States of America by
IGI Global Scientific Publishing
701 East Chocolate Avenue
Hershey, PA, 17033, USA
Tel: 717-533-8845
Fax: 717-533-7115
Website: https://www.igi-global.com E-mail: cust@igi-global.com

Library of Congress Cataloging-in-Publication Data

Names: Inobeme, Abel, 1985- editor | Mathew, John Tsado, 1983- editor | Adetunji, Charles Oluwaseun editor
Title: Advances, applications, and future perspectives of aerogels / Edited by Abel Inobeme, John Mathew, Charles Adetunji.
Description: Hershey, PA : IGI Global Scientific Publishing, [2025] | Includes bibliographical references and index. | Summary: "The aim of this book is to provide an in-depth overview and empirical findings on the chemistry of aerogels. It also discusses comprehensively the synthesis, characterization and physicochemical properties of aerogels"-- Provided by publisher.
Identifiers: LCCN 2024054544 (print) | LCCN 2024054545 (ebook) | ISBN 9798369359716 hardcover | ISBN 9798369359754 paperback | ISBN 9798369359723 ebook
Subjects: LCSH: Aerogels | Aerogels--Industrial applications | Silica gel
Classification: LCC TP862 .A38 2025 (print) | LCC TP862 (ebook)
LC record available at https://lccn.loc.gov/2024054544
LC ebook record available at https://lccn.loc.gov/2024054545

British Cataloguing in Publication Data
A Cataloguing in Publication record for this book is available from the British Library.

Table of Contents

Detailed Table of Contents

Esther Bernard, Nassarawa State University, Keffi, Nigeria
Abel Inobeme, Edo State University, Uzairue, Nigeria
Marian Agbugui, Edo State University, Uzairue, Nigeria
Jonathan Inobeme, Ahmadu Bello University, Zaria, Nigeria
Charles Oluwaseun Adetunji, Edo State University, Uzairue, Nigeria
Chinenye Anastina Eziukwu, Edo State University, Uzairue, Nigeria

Aerogels have emerged as promising materials for various environmental protection applications. Their outstanding physicochemical properties make them ideal candidates for tasks ranging from pollutant remediation to thermal insulation. In recent years, significant advances have been made in harnessing aerogels for environmental protection purposes. This paper reviews the latest developments in aerogel-based technologies, focusing on their utilization in mitigating air and water pollution, as well as in managing hazardous waste materials. Furthermore, novel strategies for enhancing the performance and durability of aerogel-based environmental solutions are discussed.

Aerogel nanocomposites, characterized by their porous and lightweight structure, exhibit a diverse array of properties and find applications across various fields. By integrating aerogels with nanomaterials, such as carbon nanotubes, graphene, or metal nanoparticles, these composites achieve enhanced mechanical strength, thermal conductivity, and electrical conductivity compared to traditional aerogels. They often display superior thermal and chemical stability, making them suitable for extreme environments. In terms of applications, aerogel nanocomposites are utilized in insulation for buildings, aerospace materials, energy storage devices, and environmental remediation technologies. Their high surface area and porosity enable efficient adsorption of pollutants from air and water. Furthermore, their lightweight nature makes them attractive for lightweight structural components in vehicles and electronics. Continued research in the synthesis methods and characterization techniques of aerogel nanocomposites is essential to unlock their full potential and expand their applications further

Aerogels have garnered significant attention in biomedical applications due to their unique properties. This review explores three key areas: biosensing, imaging, and bone regeneration. In biosensing, aerogels serve as excellent platforms for immobilizing biomolecules, enabling sensitive detection of analytes through various mechanisms such as fluorescence or surface plasmon resonance. In imaging, their low density and high surface area make aerogels promising candidates for contrast agents and drug delivery vehicles in magnetic resonance imaging and computed tomography. Furthermore, their biocompatibility and tunable porosity render aerogels suitable scaffolds for bone regeneration, promoting cell adhesion, proliferation, and mineralization. This abstract underscores the multifaceted roles aerogels play in addressing diverse biomedical challenges, paving the way for future breakthroughs in healthcare.

Aerogels have emerged as promising catalyst supports due to their high surface area, tunable porosity, and tailored surface chemistry. This review explores the catalytic application of aerogels in various reactions, including heterogeneous catalysis, photocatalysis, and enzymatic catalysis. Aerogel-based catalysts exhibit enhanced catalytic activity, selectivity, and stability compared to traditional catalysts.

The unique properties of aerogels, such as their high surface-to-volume ratio and uniform pore structure, facilitate efficient mass transfer and active site accessibility, leading to improved reaction kinetics. Moreover, the ability to functionalize aerogel surfaces enables the immobilization of catalytic species, further enhancing catalytic performance. This review highlights recent advances and future prospects in the field of aerogel-based catalysis, emphasizing the significance of aerogels as versatile and effective catalyst supports for a wide range of industrial and environmental applications

John Tsado Mathew, Department of Chemistry, Ibrahimm Badamasi Babangida Lapai, Nigeria
Jonathan Hussaini, Department of Chemistry, Ibrahim Badamasi Babangida University, Lapai, Nigeria
Yakubu Azeh, Department of Chemistry, Ibrahim Badamasi Babangida University, Lapai, Nigeria
Monday Musah, Department of Chemistry, Ibrahim Badamasi Babangida University, Lapai, Nigeria
Elijah Yanda Shaba, Department of Chemistry, Federal University of Technology, Minna, Nigeria
Ezekiel Tanko, Department of Chemistry, Federal University of Technology, Minna, Nigeria
Aishetu Ibrahim Muhammad, Department of Chemistry, Ibrahim Badamasi Babangida University, Lapai, Nigeria
Muhammad Bini Etsuyankpa, Department of Chemistry, Federal University, Lafia, Nigeria
Tanko Safiyanu Musa, Department of Chemistry, Federal University, Lafia, Nigeria
Amos Mamman, Department of Chemistry, Ibrahim Badamasi Babangida University, Lapai, Nigeria

This study evaluates the environmental and health impacts of aerogels, known for their low density, high porosity, and excellent thermal insulation. It involves a comprehensive review of literature and experimental analysis to assess factors like aerogel composition, production methods, usage scenarios, and disposal. Key environmental considerations include energy consumption, greenhouse gas emissions, and ecological toxicity. Health impacts are examined through potential exposure pathways, toxicological effects, and occupational hazards related to aerogel handling. The goal is to guide sustainable aerogel utilization by identifying and mitigating potential adverse effects on both ecosystems and human health, thereby supporting informed decision-making for safer and more responsible use of these materials.

Aerogels, with their exceptional properties including ultra-low density, high surface area, and superior thermal insulation, have garnered significant attention across various fields of research and application. In this abstract, we delve into the future perspectives of aerogel-based research, highlighting emerging trends and potential directions. One promising avenue is the advancement of aerogel manufacturing techniques, aiming for scalability, cost-effectiveness, and environmental sustainability. Additionally, the integration of aerogels into novel applications such as energy storage, catalysis, and biomedical devices presents exciting opportunities for innovation. Furthermore, exploring the potential of aerogels in emerging fields like aerospace, wearable technology, and environmental remediation holds promise for transformative breakthroughs. This abstract provides a glimpse into the future landscape of aerogel research, emphasizing the importance of continued exploration and innovation in harnessing the remarkable properties of these remarkable materials

Aerogels are a class of materials known for their exceptional physical properties, characterized by their high porosity and low density. They are composed of a network of interconnected nanoparticles or polymers, resulting in a solid with a porous structure filled with gas. This unique structure imparts several advantageous physicochemical properties to aerogels, including high surface area, low thermal conductivity, and high porosity. These properties make aerogels promising candidates for various applications, such as insulation, catalysis, drug delivery, and environmental remediation. This chapter focuses on the physicochemical properties and types of aerogels.

Preface

The field of material science has witnessed remarkable advancements over the years, leading to the development of a wide range of materials tailored for diverse applications across medicine, engineering, environmental science, and numerous other domains. Despite their significance, many conventional materials exhibit inherent limitations, necessitating the continuous search for novel and revolutionary alternatives. Aerogels have emerged as one such groundbreaking material, offering exceptional physicochemical properties that have positioned them as highly versatile and indispensable in various scientific and industrial applications.

This book, *Advances, Applications, and Future Perspectives of Aerogels*, serves as a comprehensive and up-to-date resource aimed at bridging the knowledge gap in aerogel research. It provides an in-depth exploration of the chemistry, synthesis, characterization, and physicochemical properties of aerogels. Furthermore, it highlights the recent advances and diverse applications of these extraordinary materials, spanning environmental protection, biomedical sciences, wastewater treatment, energy storage, biosensing technology, and electronics, among others. By consolidating cutting-edge research and empirical findings, this volume presents a state-of-the-art review of the most recent developments in aerogel science.

This book explores the latest advancements in aerogel research, highlighting their diverse applications in environmental protection, energy efficiency, biomedical sciences, and beyond. The following chapter overview provides insights into key developments in aerogel technology, including their role in catalysis, sensing, insulation, and sustainable development. Each chapter delves into specific aspects of aerogels, showcasing their potential to address pressing global challenges while paving the way for future innovations.

CHAPTER OVERVIEW

Chapter 1: Aerogels have emerged as promising materials for various environmental protection applications. Their outstanding physicochemical properties make them ideal candidates for pollutant remediation, thermal insulation, and hazardous waste management. This chapter reviews the latest developments in aerogel-based technologies for mitigating air and water pollution, along with novel strategies for enhancing their performance and durability.

Chapter 2: Aerogel nanocomposites, characterized by their porous and lightweight structure, exhibit enhanced mechanical strength, thermal conductivity, and electrical conductivity. By integrating aerogels with nanomaterials such as carbon nanotubes, graphene, or metal nanoparticles, these composites gain superior thermal and chemical stability. This chapter explores their applications in insulation, aerospace materials, energy storage, and environmental remediation, highlighting synthesis methods and characterization techniques.

Chapter 3: With ultra-low density, high surface area, and excellent thermal insulation, aerogels are increasingly used in sensing technologies. This chapter discusses the synthesis and functionalization of aerogels for improved sensing properties, examining their applications in detecting various analytes for environmental monitoring and biomedical diagnostics.

Chapter 4: Aerogels serve as effective sound absorbers and thermal insulators due to their nanoporous structure. This chapter explores their role in noise reduction and energy efficiency, providing insights into their applications in building materials, industrial processes, and sustainable thermal management.

Chapter 5: Aerogels play a vital role in biosensing, imaging, and bone regeneration. Their unique properties enable them to immobilize biomolecules for sensitive analyte detection, serve as contrast agents in imaging, and function as scaffolds for bone tissue engineering. This chapter reviews the diverse biomedical applications of aerogels and their potential for future healthcare innovations.

Chapter 6: Aerogels provide high surface area, tunable porosity, and tailored surface chemistry, making them effective catalyst supports. This chapter explores their role in heterogeneous catalysis, photocatalysis, and enzymatic catalysis, highlighting recent advances in aerogel-based catalysts for industrial and environmental applications.

Chapter 7: This chapter evaluates the environmental and health impacts of aerogels, including their composition, production, usage, and disposal. It examines key factors such as energy consumption, greenhouse gas emissions, and potential toxicological effects, offering guidance on sustainable aerogel utilization for safer and more responsible applications.

Chapter 8: This chapter delves into emerging trends and future directions in aerogel research, focusing on scalable manufacturing techniques, cost-effectiveness, and environmental sustainability. It also explores novel applications in energy storage, catalysis, biomedical devices, aerospace, wearable technology, and environmental remediation.

Chapter 9: Aerogels possess exceptional physical properties, including high porosity, low density, and high surface area. This chapter provides an in-depth discussion of the physicochemical properties and classification of aerogels, analyzing their potential across diverse fields such as insulation, catalysis, drug delivery, and environmental remediation.

The motivation behind this edited collection is to offer a well-structured and accessible reference for professionals, researchers, engineers, and students who seek to deepen their understanding of aerogels and their transformative potential. This book is designed to serve as an invaluable tool for those engaged in the development and application of novel materials across various scientific and technological fields. We believe that the insights provided herein will inspire further research and innovation in aerogel science, driving the field towards new frontiers of discovery and practical implementation.

We extend our heartfelt appreciation to all contributors whose expertise and dedication have made this volume possible. We also acknowledge the invaluable support of our respective institutions in facilitating this endeavor. It is our hope that this book will serve as a guiding resource for both academia and industry, fostering continued advancements in the fascinating world of aerogels.

Abel Inobeme

Edo University, Uzairue, Nigeria

John Tsado Mathew

Ibrahimm Badamasi Babangida Lapai, Nigeria

Charles Oluwaseun Adetunji

Edo University, Auchi, Nigeria

Chapter 1
Advances in the Use of Aerogel for Environmental Protection

Esther Bernard
Nassarawa State University, Keffi, Nigeria

Jonathan Inobeme
Ahmadu Bello University, Zaria, Nigeria

Abel Inobeme
Edo State University, Uzairue, Nigeria

Charles Oluwaseun Adetunji
Edo State University, Uzairue, Nigeria

Marian Agbugui
Edo State University, Uzairue, Nigeria

Chinenye Anastina Eziukwu
Edo State University, Uzairue, Nigeria

ABSTRACT

Aerogels have emerged as promising materials for various environmental protection applications. Their outstanding physicochemical properties make them ideal candidates for tasks ranging from pollutant remediation to thermal insulation. In recent years, significant advances have been made in harnessing aerogels for environmental protection purposes. This paper reviews the latest developments in aerogel-based technologies, focusing

DOI: 10.4018/979-8-3693-5971-6.ch001

on their utilization in mitigating air and water pollution, as well as in managing hazardous waste materials. Furthermore, novel strategies for enhancing the performance and durability of aerogel-based environmental solutions are discussed.

9.1 INTRODUCTION

Aerogels, often referred to as "frozen smoke" or "solid smoke," are remarkable materials characterized by their ultra-lightweight, high porosity, and extensive surface area. Originally developed in the 1930s by Samuel Kistler, aerogels have since captivated the attention of scientists and engineers due to their unique properties and wide-ranging applications. One of the most promising and rapidly expanding areas of aerogel research is their utilization in environmental protection.

The imperative for environmental protection has never been more pressing. With escalating levels of pollution, habitat destruction, and climate change, there is an urgent need for innovative solutions to mitigate environmental degradation and safeguard the planet's ecosystems. In this context, aerogels have emerged as valuable tools for addressing a myriad of environmental challenges, owing to their versatile properties and customizable characteristics.

Aerogels are composed of a network of interconnected nanoparticles, resulting in a solid yet highly porous structure. This structure imparts several advantageous properties to aerogels, including exceptionally low density, typically ranging from 0.001 to 0.5 g/cm^3, making them some of the lightest solid materials known to humankind. Furthermore, aerogels exhibit high porosity, with pore sizes ranging from nanometers to micrometers, and possess an enormous internal surface area, often exceeding 1000 square meters per gram. These features enable aerogels to effectively capture, adsorb, or catalytically convert various pollutants, making them ideal candidates for environmental protection applications (Nita et al. 2020).

The utilization of aerogels for environmental protection is multifaceted, encompassing air and water pollution remediation, as well as hazardous waste management (figure 1). In the realm of air pollution, aerogels have demonstrated remarkable efficacy in adsorbing harmful gases and particulate matter, thereby improving air quality and reducing human health risks. Additionally, aerogels have been employed in catalytic conversion processes for the degradation of volatile organic compounds (VOCs) and other airborne pollutants, facilitating their conversion into less harmful substances (Franco et al. 2021).

In the domain of water pollution mitigation, aerogels play a crucial role in the removal of contaminants from aqueous environments. Their high surface area and porosity enable efficient absorption of pollutants such as heavy metals, organic compounds, and microorganisms, thereby purifying water resources and safeguarding ecosystems. Furthermore, aerogel-based filtration technologies offer a sustainable and cost-effective approach to water treatment, with applications ranging from municipal wastewater treatment to industrial effluent remediation (Ahmad *et al.* 2023).

Figure 1. Application of aerogels in environment protection

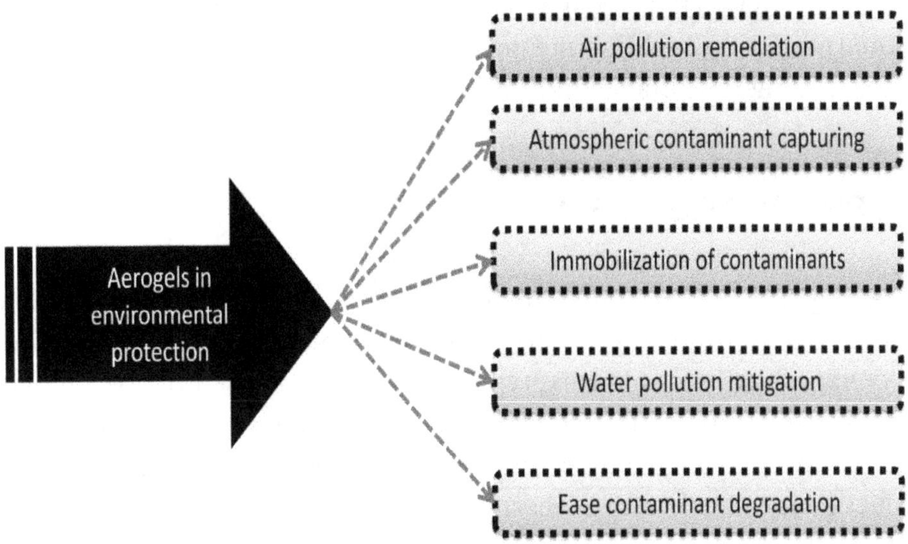

Moreover, aerogels are increasingly being utilized in hazardous waste management to encapsulate and immobilize toxic substances, preventing their release into the environment and mitigating risks to human health and ecological integrity. Whether used for the remediation of contaminated sites or the containment of hazardous materials, aerogels offer a versatile and environmentally benign solution to the challenges posed by hazardous waste disposal (Boccia et al. 2023).

In recent years, significant advances have been made in harnessing the potential of aerogels for environmental protection purposes. Researchers have focused on enhancing the performance and durability of aerogel-based systems through innovative approaches such as surface modifications and composite formulations. These advancements have yielded aerogel materials with tailored properties and enhanced capabilities, further expanding their utility in environmental applications (Bakhori et al. 2023).

Overall, the continued exploration and refinement of aerogel technologies hold great promise for addressing pressing environmental challenges and promoting sustainable development. By leveraging the unique properties of aerogels, researchers and engineers can develop effective solutions for pollution remediation, water purification, and hazardous waste management, contributing to a healthier and more resilient planet. In this review, we will delve into the recent advances in the use of aerogels for environmental protection, highlighting their applications, properties, and potential for mitigating environmental threats (Annu et al. 2024).

9.2 PROPERTIES OF AEROGELS

Aerogels possess several remarkable properties that make them highly desirable for a wide range of applications (**Table 1**). Firstly, they exhibit an extraordinarily low density, typically ranging from 0.001 to 0.5 g/cm^3, making them some of the lightest solid materials known. This low density enables aerogels to be utilized in lightweight applications where weight reduction is critical, such as thermal insulation in aerospace components (Khan et al.2023).

Secondly, aerogels are characterized by their high porosity, featuring a network of interconnected nanoparticle structures. This porosity provides aerogels with an immense internal surface area, often exceeding 1000 square meters per gram. As a result, aerogels offer exceptional adsorption capabilities, making them effective for capturing pollutants from air or water, as well as for catalyzing chemical reactions.

Furthermore, aerogels demonstrate excellent thermal insulation properties due to their low thermal conductivity. This makes them ideal for applications where temperature regulation and energy conservation are essential, such as in building insulation or cryogenic storage. Overall, the unique combination of low density, high porosity, large surface area, and thermal insulation make aerogels versatile materials with significant potential for various environmental and technological applications (Franco et al. 2021).

Table 1. Main properties of aerogels

Property	Description	Importance/Relevance	Examples/Applications	References
Low Density	Aerogels have extremely low density, typically ranging from 0.001 to 0.5 g/cm^3.	Facilitates lightweight applications, reduces material usage, and enables efficient transportability.	Thermal insulation, lightweight structural components, aerospace materials.	Jin et al. 2023.
High Porosity	Aerogels possess a highly porous structure with interconnected nanoparticle networks.	Provides a large surface area for adsorption and enhances permeability for gas and liquid transport.	Adsorption of pollutants, catalytic reactions, filtration, and absorption processes.	Shah et al. 2021.
Large Surface Area	Aerogels exhibit an enormous internal surface area, often exceeding 1000 square meters per gram.	Maximizes interaction with gases and liquids, enhances adsorption capacity, and promotes efficient reactions.	Adsorption of pollutants, catalytic conversions, energy storage, and sensing.	Budtova et al. 2020.
Thermal Insulation	Aerogels demonstrate excellent thermal insulation properties due to their low thermal conductivity.	Enables energy conservation, temperature regulation, and improved efficiency in various applications.	Building insulation, protective clothing, cryogenic storage, and thermal barriers.	Wu et al. 2023.

9.3 APPLICATIONS OF AEROGELS IN ENVIRONMENTAL PROTECTION

Aerogels have emerged as valuable tools in the realm of environmental protection due to their unique properties and versatile applications. From mitigating air and water pollution to managing hazardous waste, aerogels offer innovative solutions to pressing environmental challenges (Romero-Montero et al. 2023). Aerogels have unique properties that make them suitable in environmental protection (figure 2).

One of the primary applications of aerogels in environmental protection is in air pollution remediation. Aerogels are effective at capturing airborne pollutants such as particulate matter, volatile organic compounds (VOCs), and harmful gases. Their high surface area and porosity enable efficient adsorption of pollutants, thereby improving air quality and reducing health risks for humans and ecosystems. Aerogel-based filters can be deployed in industrial facilities, vehicles, and indoor environments to remove pollutants before they are released into the atmosphere (Drdova et al. 2022).

In addition to adsorption, aerogels can also catalyze chemical reactions to degrade airborne pollutants. Functionalized aerogels can act as catalysts for the conversion of harmful gases into less toxic compounds through processes such as oxidation or reduction. This catalytic activity makes aerogels valuable in combating air pollution from sources such as vehicle emissions, industrial exhausts, and urban smog (AlSoubaihi et al. 2023).

Aerogels are also instrumental in water pollution mitigation efforts. Their high porosity and large surface area make them excellent adsorbents for contaminants present in water bodies, including heavy metals, organic pollutants, and microorganisms. Aerogel-based filtration systems can effectively remove these contaminants, purifying water for various applications such as drinking, agriculture, and industrial processes (Lamy-Mendes et al. 2023).

Moreover, aerogels can be tailored to target specific pollutants through surface modifications and functionalization. By introducing specific chemical groups or nanoparticles onto the aerogel surface, researchers can enhance its affinity for particular pollutants, improving its effectiveness in water purification applications. This versatility allows aerogels to address a wide range of water pollution challenges, from industrial wastewater treatment to the remediation of contaminated groundwater sites (Lamy-Mendesetal.2023).

Furthermore, aerogels play a crucial role in hazardous waste management. They can be used to encapsulate and immobilize toxic substances, preventing their release into the environment and minimizing risks to human health and ecosystems. Aerogel matrices can trap hazardous materials, preventing their migration and leaching into soil or groundwater. This containment strategy is particularly useful for the remediation of contaminated sites, such as brownfields or industrial landfills, where hazardous waste poses significant environmental threats (Ahmad *et al.* 2023).

Additionally, aerogels offer advantages in terms of ease of handling and disposal compared to traditional hazardous waste containment materials. Their lightweight nature and low volume facilitate transportation and storage, reducing logistical challenges and costs associated with waste management. Furthermore, aerogels can be designed to degrade or transform hazardous compounds over time, offering sustainable solutions for long-term waste remediation (Nguyen et al. 2022).

In recent years, significant advancements have been made in the development of aerogel-based environmental protection technologies. Researchers are exploring novel materials and fabrication techniques to enhance the performance and efficiency of aerogels in pollution control and waste management applications. For example, the integration of nanoparticles or functional polymers into aerogel matrices can improve adsorption capacity, selectivity, and durability (Bakhori et al. 2023).

Furthermore, efforts are underway to scale up aerogel production and reduce manufacturing costs to make these technologies more accessible for widespread deployment. Innovations in aerogel synthesis, such as supercritical drying methods and continuous manufacturing processes, are driving progress towards commercialization and large-scale implementation (Lebedevet al. 2021).

Overall, aerogels hold great promise for addressing environmental challenges and promoting sustainability. Their unique properties and versatile applications make them valuable tools in pollution control, water purification, and hazardous waste management efforts. As research continues to advance, aerogel-based technologies are expected to play an increasingly important role in safeguarding the environment and preserving natural resources for future generations (Khan et al. 2023).

Figure 2. Properties of aerogels useful in environmental remediation

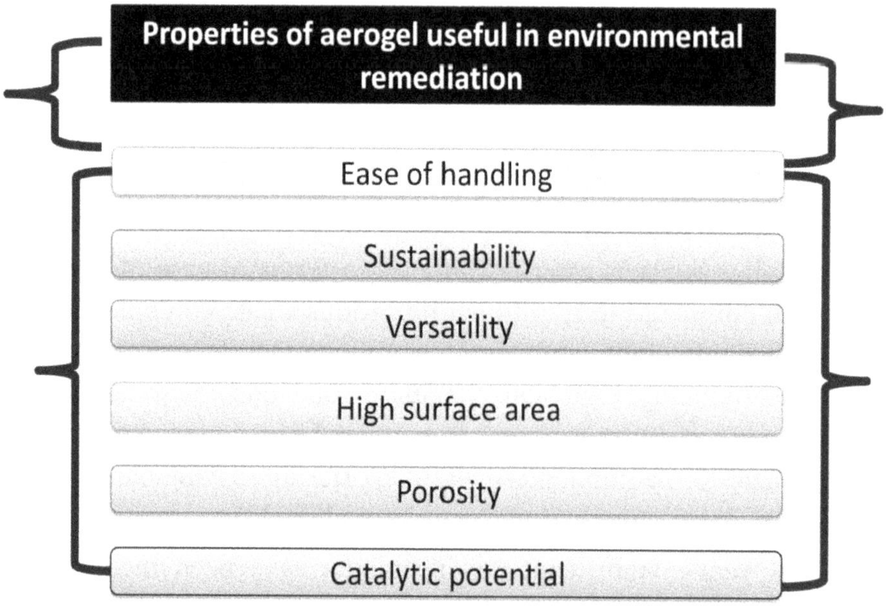

9.4 WATER POLLUTION MITIGATION

Water pollution is a pressing global environmental challenge that poses significant risks to human health, ecosystems, and biodiversity. Sources of water pollution include industrial effluents, agricultural runoff, urban sewage, and improper waste disposal. To address these issues, innovative solutions are needed for water pollution mitigation, and aerogels offer promising opportunities in this regard (Bashir et al. 2020).

Aerogels, with their high surface area, porous structure, and customizable properties, are effective tools for removing contaminants from water bodies. One of the primary methods of water pollution mitigation using aerogels is through adsorption. Aerogels can selectively adsorb a wide range of pollutants, including heavy metals, organic compounds, and microorganisms, from aqueous solutions. This adsorption process relies on the interaction between the pollutant molecules and the surface functional groups or pores of the aerogel. The high porosity of aerogels provides a large surface area for pollutant adsorption, enhancing their efficiency in water treatment applications. Additionally, aerogels can be tailored to target specific contaminants by modifying their surface chemistry. Functionalizing aerogels with specific chemical groups or nanoparticles can increase their affinity for particular pollutants, improving their adsorption capacity and selectivity (Franco et al. 2021).

Another method of water pollution mitigation using aerogels is through filtration. Aerogel-based filtration systems can remove suspended solids, colloids, and other particulate matter from water, thereby improving clarity and reducing turbidity. The porous structure of aerogels allows water to flow through while trapping contaminants, providing an effective means of purifying water for various uses. Furthermore, aerogels can be incorporated into composite materials for enhanced water purification capabilities. By combining aerogels with other materials such as membranes or granular media, researchers can create multifunctional systems with improved adsorption, filtration, and separation properties. These aerogel-based composites offer versatility and efficiency in addressing complex water pollution challenges. Aerogels are also valuable in addressing emerging contaminants and micropollutants in water. These include pharmaceuticals, personal care products, pesticides, and industrial chemicals, which are not effectively removed by conventional water treatment processes. Aerogels can selectively adsorb these micropollutants due to their high surface area and tailored surface chemistry, offering a solution to the growing concern of water contamination by emerging pollutants (Romero-Montero et al. 2023).

Moreover, aerogels can be regenerated and reused multiple times, making them cost-effective and sustainable solutions for water pollution mitigation. After adsorbing pollutants, aerogels can be regenerated by desorbing the contaminants using techniques such as thermal treatment or solvent extraction. This allows the aerogels to be reused, reducing the need for frequent replacement and minimizing waste generation. In addition to adsorption and filtration, aerogels can also be used for the catalytic degradation of water pollutants. Functionalized aerogels can serve as catalysts for chemical reactions that break down harmful contaminants into less toxic or inert substances. This catalytic activity can be employed in advanced oxidation processes (AOPs) or photocatalytic reactions for the degradation of organic pollutants, disinfection of microorganisms, and removal of recalcitrant compounds from water (Franco et al, 2021).

The application of aerogels in water pollution mitigation extends beyond traditional treatment methods to include innovative approaches such as floating aerogel-based systems for the remediation of surface water bodies. These floating platforms can absorb pollutants directly from the water surface, targeting contaminants such as oil spills, floating debris, and surface-active chemicals. By harnessing the buoyancy and adsorption capacity of aerogels, these systems offer efficient and environmentally friendly solutions for cleaning up polluted water bodies (Li et al. 2023).

In general, aerogels hold great promise for mitigating water pollution through adsorption, filtration, catalysis, and innovative remediation technologies. Their unique properties and versatility make them valuable tools for removing contaminants from water bodies, improving water quality, and safeguarding human health and the environment. As research and development in aerogel-based water treatment technologies continue to advance, these materials are expected to play an increasingly important role in addressing global water pollution challenges (Khan et al. 2023).

9.5 ENCAPSULATION AND CONTAINMENT

Encapsulation and containment are critical strategies in hazardous waste management aimed at preventing the release of toxic substances into the environment and minimizing risks to human health and ecosystems. Aerogels offer innovative solutions for encapsulating and containing hazardous materials, providing effective barriers against migration and leaching. One of the primary advantages of using aerogels for encapsulation is their high surface area and porosity, which allow them to trap and immobilize toxic substances within their porous structure. Hazardous materials such as heavy metals, radioactive isotopes, and organic pollutants can be encapsulated within aerogels through physical entrapment or chemical bonding, preventing their dispersion into the surrounding environment (Lee et al. 2023).

The porous nature of aerogels enables them to act as effective barriers against the migration of hazardous contaminants. When encapsulated within aerogels, toxic substances are confined within the material matrix, reducing the risk of leaching into soil, groundwater, or surface water. This containment strategy is particularly beneficial for the remediation of contaminated sites, such as industrial landfills, brownfields, and mining sites, where hazardous waste poses significant environmental threats. Moreover, aerogels can be engineered to selectively capture specific contaminants, enhancing their effectiveness in encapsulation and containment applications. Functionalized aerogels with tailored surface chemistry can exhibit high affinity for target pollutants, allowing for the selective removal and immobilization of hazardous materials from contaminated environments. This targeted encapsulation approach minimizes the volume of waste generated and facilitates the remediation of complex contamination scenarios (Romero-Montero et al. 2023).

In addition to encapsulating hazardous materials, aerogels can also serve as protective barriers for the containment of radioactive waste. Aerogel-based matrices can encapsulate radioactive isotopes, such as uranium, plutonium, and cesium, preventing their dispersion and minimizing radiation exposure risks. The high porosity and thermal stability of aerogels make them suitable candidates for long-term storage and disposal of radioactive waste, offering enhanced safety and security measures for nuclear facilities and waste repositories. Furthermore, aerogels offer advantages in terms of ease of handling and disposal compared to traditional containment materials. Their lightweight nature and low volume facilitate transportation, installation, and maintenance, reducing logistical challenges and costs associated with hazardous waste management. Aerogels can be easily molded into various shapes and sizes to accommodate specific containment requirements, providing flexibility in deployment and application (Asmussen et al. 2022).

Another benefit of using aerogels for encapsulation and containment is their ability to encapsulate multiple contaminants simultaneously. Mixed waste streams containing a combination of hazardous substances can be encapsulated within aerogels, allowing for comprehensive remediation of contaminated sites with diverse pollutant profiles. This versatility enables aerogels to address complex contamination scenarios and provides a holistic approach to hazardous waste management. Moreover, aerogels can be designed to undergo degradation or transformation over time, offering sustainable solutions for long-term waste remediation. Biodegradable aerogels can break down into non-toxic byproducts through natural processes, minimizing the need for post-remediation cleanup and reducing environmental impact. This environmentally friendly approach aligns with the principles of green chemistry and sustainable development, promoting responsible stewardship of natural resources and ecosystems (Franco et al. 2021; Mathew et al., 2024a).

In addition, aerogels offer innovative solutions for encapsulating and containing hazardous materials in environmental remediation efforts. Their high surface area, porosity, and customizable properties make them effective barriers against the migration and dispersion of toxic substances, providing enhanced safety and security measures for hazardous waste management. As research and development in aerogel technologies continue to advance, these materials are expected to play an increasingly important role in safeguarding human health and the environment from the risks associated with hazardous waste contamination (Bakhori et al. 2023).

9.6 REMEDIATION OF CONTAMINATED SITES

Remediation of contaminated sites is a crucial aspect of environmental protection, aimed at restoring land and water resources affected by pollutants and hazardous materials. Aerogels offer innovative solutions for remediation efforts, providing effective tools for capturing, immobilizing, and removing contaminants from contaminated environments. One of the primary challenges in the remediation of contaminated sites is the presence of persistent organic pollutants (POPs), such as polychlorinated biphenyls (PCBs), polycyclic aromatic hydrocarbons (PAHs), and chlorinated solvents. These pollutants can persist in the environment for extended periods, posing risks to human health and ecosystems. Aerogels can be employed to remediate sites contaminated with POPs through adsorption, encapsulation, and degradation processes (Franco et al.2021).

Aerogels' high surface area and porosity make them effective adsorbents for capturing POPs from soil, groundwater, and surface water. Aerogel-based adsorbents can selectively trap organic pollutants, removing them from the environment and reducing their bioavailability and toxicity. This adsorption process mitigates the risks associated with POPs exposure and facilitates the restoration of contaminated sites to their natural state. Furthermore, aerogels can encapsulate POPs within their porous structure, preventing their migration and leaching into the surrounding environment. Encapsulated contaminants are immobilized within the aerogel matrix, reducing their mobility and exposure risks. This containment strategy offers long-term stability and security for contaminated sites, minimizing the potential for recontamination and ensuring environmental protection (Boccia et al., 2024).

In addition to adsorption and encapsulation, aerogels can catalyze chemical reactions for the degradation of organic pollutants. Functionalized aerogels with catalytic properties can facilitate the decomposition of POPs into less toxic or inert substances through oxidation, reduction, or hydrolysis processes. This catalytic degradation approach offers an efficient and sustainable means of remediation, promoting the transformation of hazardous contaminants into harmless byproducts. Moreover, aerogels can be utilized in ex-situ remediation techniques for the removal of contaminants from soil and groundwater. Aerogel-based adsorbents can be applied directly to contaminated soil or introduced into groundwater wells to capture and remove pollutants through adsorption or filtration processes. This approach allows for targeted treatment of contaminated media, minimizing disturbance to surrounding ecosystems and reducing remediation costs (Li et al., 2023).

Furthermore, aerogels can be integrated into barrier systems for the containment and isolation of contaminated areas. Aerogel-based barriers can be installed around contaminated sites to prevent the migration of pollutants and protect adjacent areas from contamination. These barriers act as impermeable barriers, blocking the flow of contaminated groundwater, surface water, or leachate, and providing a physical barrier against the spread of pollutants. Additionally, aerogels offer advantages in terms of environmental compatibility and sustainability compared to traditional remediation materials. Aerogels are lightweight, non-toxic, and biocompatible, minimizing their environmental impact and reducing risks to human health and ecosystems. Their high surface area and porosity allow for efficient pollutant capture and removal, while their customizable properties enable tailored solutions for specific remediation challenges (Cao et al., 2021; Mathew et al., 2023a).

Generally, aerogels offer innovative and effective solutions for the remediation of contaminated sites, providing versatile tools for capturing, immobilizing, and degrading pollutants. Their high surface area, porosity, and customizable properties make them ideal materials for addressing a wide range of environmental contamination scenarios. As research and development in aerogel technologies continue to advance, these materials are expected to play an increasingly important role in restoring contaminated environments and protecting human health and ecosystems from the risks associated with pollution (Boccia et al., 2023).

9.7 RECENT ADVANCES IN AEROGEL TECHNOLOGIES

Recent advances in aerogel technologies have led to significant improvements in their properties, performance, and applicability (**Table 2**). One key area of advancement is surface modifications, where researchers are exploring novel chemical and physical treatments to tailor aerogel surfaces for specific applications. These modifications enhance adsorption capacity, selectivity, and stability, enabling aerogels to address diverse environmental, industrial, and technological challenges (Khan et al. 2023).

Additionally, the development of composite formulations has expanded the range of aerogel-based materials, combining aerogels with other substances to create multifunctional composites with enhanced properties. These composites exhibit improved mechanical strength, thermal stability, and durability, opening up new possibilities for aerogel applications in fields such as construction, aerospace, and energy storage. Furthermore, efforts to enhance the durability of aerogels have resulted in innovative strategies to improve their structural integrity and longevity. By addressing challenges related to brittleness, degradation, and mechanical weakness, researchers are ensuring the long-term performance of aerogel-based materials in demanding environments (Lázár et al. 2023).

Moreover, advancements in continuous manufacturing processes are making aerogel production more scalable, cost-effective, and accessible. Continuous drying techniques, roll-to-roll processing, and automation are streamlining the manufacturing process, paving the way for large-scale production and widespread deployment of aerogel technologies. Overall, these recent advances in aerogel technologies are driving progress towards innovative solutions for a wide range of applications, from environmental remediation to advanced materials development.

Table 2. Recent advances in aerogel technologies

Category	Description	Importance/Relevance	Examples/Advancements	References
Surface Modifications	Chemical or physical alterations to aerogel surfaces to enhance specific properties or functionalities.	Improves adsorption capacity, selectivity, and stability; enables tailored solutions for diverse applications.	Introduction of functional groups, surface coatings, and nanostructures for enhanced performance.	Lamy-Mendes et al. 2019; Mathew et al., 2024b
Composite Formulations	Integration of aerogels with other materials to create multifunctional composites with enhanced properties.	Enhances mechanical strength, thermal stability, and durability; expands range of applications and functionalities.	Incorporation of aerogels into polymers, ceramics, metals, and nanocomposites for diverse uses.	Bakhori et al. 2023.
Durability Enhancements	Strategies to improve the structural integrity and longevity of aerogel-based materials.	Addresses challenges related to brittleness, degradation, and mechanical weakness; ensures long-term performance.	Cross-linking, reinforcement, and surface modifications to enhance stability and robustness.	Ghica et al. 2023.
Continuous Manufacturing Processes	Development of scalable and cost-effective methods for aerogel production on an industrial scale.	Facilitates mass production, reduces manufacturing costs, and increases accessibility of aerogel technologies.	Continuous drying techniques, roll-to-roll processing, and automation for large-scale manufacturing.	Singh et al. 2022.

9.8 REAL-WORLD APPLICATIONS OF AEROGEL-BASED ENVIRONMENTAL SOLUTIONS

Aerogel-based environmental solutions have found practical applications in various real-world scenarios, addressing pressing environmental challenges and promoting sustainability. These innovative solutions leverage the unique properties of aerogels to mitigate pollution, conserve resources, and protect ecosystems. Here are some notable examples of aerogel-based environmental applications (Noman et al. 2021; Mathew et al., 2023b).

Oil Spill Remediation: Aerogels have been utilized in the cleanup of oil spills, providing an effective means of containing and recovering spilled oil from water surfaces. Aerogel-based sorbent materials can selectively adsorb hydrocarbons while repelling water, enabling the efficient removal of oil from contaminated water bodies. These aerogel-based sorbents offer advantages such as high adsorption capacity, rapid uptake rates, and reusability, making them valuable tools for mitigating the environmental impacts of oil spills (Elhenawy et al. 2021).

Air Pollution Control: Aerogels are employed in air filtration systems for the removal of particulate matter, volatile organic compounds (VOCs), and other airborne pollutants. Aerogel-based filters exhibit high surface area and porosity, allowing them to effectively capture pollutants from air streams. These filters can be integrated into HVAC systems, industrial exhausts, and indoor air purifiers to improve air quality and reduce human exposure to harmful pollutants. Additionally, aerogel-based catalytic converters are used for the degradation of VOCs and other pollutants through catalytic reactions, further enhancing air pollution control measures (Zhou et al. 2021; Inobeme et al., 2023).

Water Purification: Aerogels play a crucial role in water purification technologies, offering efficient solutions for the removal of contaminants from aqueous environments. Aerogel-based adsorbents and filtration systems can capture heavy metals, organic pollutants, and microorganisms from water sources, thereby enhancing water quality and safeguarding public health. These aerogel-based water purification systems are deployed in municipal water treatment plants, industrial wastewater facilities, and decentralized drinking water systems, providing reliable and sustainable solutions for clean water supply (Romero-Montero et al. 2023).

Soil Remediation: Aerogels are utilized in soil remediation efforts to mitigate contamination from hazardous substances such as heavy metals, pesticides, and industrial chemicals. Aerogel-based materials can encapsulate and immobilize contaminants within soil matrices, preventing their migration and reducing risks to human health and ecosystems. Additionally, aerogel-based adsorbents can be applied to soil surfaces or introduced into groundwater wells to capture and remove pollutants through adsorption processes, facilitating the restoration of contaminated soils and groundwater resources (Priya et al. 2023).

Greenhouse Gas Capture: Aerogels are being explored for carbon capture and storage (CCS) applications, aiming to mitigate greenhouse gas emissions and combat climate change. Aerogel-based sorbents can selectively adsorb carbon dioxide (CO_2) from industrial flue gases and other emission sources, enabling the capture and sequestration of CO_2 for long-term storage or utilization. These aerogel-based CCS technologies offer potential solutions for reducing CO_2 emissions from power plants, industrial facilities, and other sources, contributing to global efforts to mitigate climate change and transition towards a low-carbon economy (Amaraweera et al. 2023).

Energy Storage and Conversion: Aerogels are utilized in energy storage and conversion devices such as batteries, supercapacitors, and fuel cells. Aerogel-based electrodes and electrolytes offer high surface area, porosity, and electrical conductivity, enhancing the performance and efficiency of energy storage and conversion systems. These aerogel-based technologies enable the development of lightweight, compact, and high-performance energy storage devices for applications in renewable energy systems, electric vehicles, and portable electronics (Kausar et al. 2023).

Overall, aerogel-based environmental solutions demonstrate significant potential for addressing a wide range of environmental challenges and promoting sustainable development. By harnessing the unique properties of aerogels, these innovative solutions offer efficient, cost-effective, and environmentally friendly approaches to pollution control, resource conservation, and ecosystem protection. Continued research and development in aerogel technologies are expected to further advance their applications and contribute to global efforts towards a cleaner, healthier, and more sustainable environment (Noman et al. 2021; Inobeme et al., 2024).

9.9 CONCLUSION

The advances in the use of aerogels for environmental protection represent a significant stride towards sustainable development and the preservation of our planet's ecosystems. With their remarkable properties such as high surface area, porosity, and versatility, aerogels offer innovative solutions to mitigate pollution, conserve resources, and safeguard human health and the environment. From oil spill remediation to air and water purification, aerogel-based technologies have demonstrated their effectiveness in addressing a wide range of environmental challenges. Recent advancements in aerogel surface modifications, composite formulations, and manufacturing processes have further enhanced their performance and applicability in real-world scenarios.

As we continue to face pressing environmental issues such as climate change, pollution, and habitat degradation, the importance of aerogels in environmental protection cannot be overstated. By leveraging the unique properties of aerogels and fostering interdisciplinary collaboration, we can harness the full potential of these materials to develop sustainable solutions for a cleaner, healthier planet. Moving forward, continued research, innovation, and investment in aerogel technologies are essential to further unlock their capabilities and maximize their impact on environmental protection. By embracing aerogels as valuable tools in our environmental arsenal, we can work towards a more resilient and sustainable future for generations to come.

REFERENCES

Ahmad, A., & Kamaruddin, M. A., H P S, A. K., Yahya, E. B., Muhammad, S., Rizal, S., Ahmad, M. I., Surya, I., & Abdullah, C. K. (. (2023). Recent Advances in Nanocellulose Aerogels for Efficient Heavy Metal and Dye Removal. *Gels (Basel, Switzerland)*, *9*(5), 416. DOI: 10.3390/gels9050416 PMID: 37233007

Al Soubaihi, R. M., Saoud, K. M., Awadallah-F, A., Elkhatat, A. M., Al-Muhtaseb, S. A., & Dutta, J. (2023). Investigation of palladium catalysts in mesoporous silica support for CO oxidation and CO2 adsorption. *Heliyon*, *9*(7), e18354. DOI: 10.1016/j.heliyon.2023.e18354 PMID: 37539214

Amaraweera, S. M., Gunathilake, C. A., Gunawardene, O. H. P., Dassanayake, R. S., Cho, E. B., & Du, Y. (2023). Carbon Capture Using Porous Silica Materials. *Nanomaterials (Basel, Switzerland)*, *13*(14), 2050. DOI: 10.3390/nano13142050 PMID: 37513061

Annu, M., Mittal, M., Tripathi, S., & Shin, D. K. (2024). BiopolymericNanocomposites for Wastewater Remediation: An Overview on Recent Progress and Challenges. *Polymers*, *16*(2), 294. DOI: 10.3390/polym16020294 PMID: 38276702

Asmussen, R. M., Turner, J., Chong, S., & Riley, B. J. (2022). Review of recent developments in iodine wasteform production. *Frontiers in Chemistry*, *10*, 1043653. DOI: 10.3389/fchem.2022.1043653 PMID: 36618856

Bakhori, N. M., Ismail, Z., Hassan, M. Z., & Dolah, R. (2023). Emerging Trends in Nanotechnology: Aerogel-Based Materials for Biomedical Applications. *Nanomaterials (Basel, Switzerland)*, *13*(6), 1063. DOI: 10.3390/nano13061063 PMID: 36985957

Bashir, I., Lone, F. A., Bhat, R. A., Mir, S. A., Dar, Z. A., & Dar, S. A. (2020). Concerns and Threats of Contamination on Aquatic Ecosystems. Bioremediation and Biotechnology: Sustainable Approaches to Pollution Degradation, 1–26. DOI: 10.1007/978-3-030-35691-0_1

Boccia, A. C., Neagu, M., & Pulvirenti, A. (2023). Bio-Based Aerogels for the Removal of Heavy Metal Ions and Oils from Water: Novel Solutions for Environmental Remediation. *Gels (Basel, Switzerland), 10*(1), 32. DOI: 10.3390/gels10010032 PMID: 38247754

Boccia, A. C., Pulvirenti, A., García-González, C. A., Grisi, F., & Neagu, M. (2023). Compendium of Safety Regulatory for Safe Applications of Aerogels. *Gels (Basel, Switzerland), 9*(11), 842. DOI: 10.3390/gels9110842 PMID: 37998932

Budtova, T., Aguilera, D. A., Beluns, S., Berglund, L., Chartier, C., Espinosa, E., Gaidukovs, S., Klimek-Kopyra, A., Kmita, A., Lachowicz, D., Liebner, F., Platnieks, O., Rodríguez, A., Navarro, L. K. T., Zou, F., & Buwalda, S. J. (2020). Biorefinery Approach for Aerogels. *Polymers, 12*(12), 2779. DOI: 10.3390/polym12122779 PMID: 33255498

Cao, B., Xu, J., Wang, F., Zhang, Y., & O'Connor, D. (2021). Vertical Barriers for Land Contamination Containment: A Review. *International Journal of Environmental Research and Public Health, 18*(23), 12643. DOI: 10.3390/ijerph182312643 PMID: 34886373

Drdova, S., Zhao, S., Giannakou, M., Sivaraman, D., Guerrero-Alburquerque, N., Bonnin, A., Pauer, R., Pan, Z., Billeter, E., Siqueira, G., Zeng, Z., Koebel, M. M., Malfait, W. J., & Wang, J. (2022). Biomimetic Light-Driven Aerogel Passive Pump for Volatile Organic Pollutant Removal. Advanced science (Weinheim, Baden-Wurttemberg, Germany), 9(11), e2105819. DOI: 10.1002/advs.202105819

Elhenawy, S., Khraisheh, M., AlMomani, F., Hassan, M. K., Al-Ghouti, M. A., & Selvaraj, R. (2021). Recent Developments and Advancements in Graphene-Based Technologies for Oil Spill Cleanup and Oil-Water Separation Processes. *Nanomaterials (Basel, Switzerland), 12*(1), 87. DOI: 10.3390/nano12010087 PMID: 35010035

Franco, P., Cardea, S., Tabernero, A., & De Marco, I. (2021). Porous Aerogels and Adsorption of Pollutants from Water and Air: A Review. *Molecules (Basel, Switzerland), 26*(15), 4440. DOI: 10.3390/molecules26154440 PMID: 34361593

Ghica, M. E., Mandinga, J. G. S., Linhares, T., Almeida, C. M. R., & Durães, L. (2023). Improvement of the Mechanical Properties of Silica Aerogels for Thermal Insulation Applications through a Combination of Aramid Nanofibres and Microfibres. *Gels (Basel, Switzerland)*, *9*(7), 535. DOI: 10.3390/gels9070535 PMID: 37504414

Inobeme, A., Adetunji, C. O., Mathew, J T., Ajai, A. I., Inobeme, J., Bamigboye, M. O., Onyeaku, S., Maliki, M., Eziukwu, C. &Tawa, K. (2023). Nanotechnology for Bioremediation of Heavy Metals. In book: Microbial Technologies in Industrial Wastewater Treatment. Springer Nature Singapore, 19-30. .DOI: 10.1007/978-981-99-2435-6_2

Inobeme, A., Mathew, J.T., Adetunji, C.O., Agbugui, M.O., Osarenren, E.J., Oti, C., Ngonso, B.F., Inobeme, J., Oyedolapo, M.B., Bernard, E. &Jibrin, N.A. (2024). Biocomposites with graphene derivatives. In book: Advances in Biocomposites and their Applications, Elsevier Inc., 149-166. DOI: DOI: 10.1016/B978-0-443-19074-2.00005-8

Jin, R., Zhou, Z., Liu, J., Shi, B., Zhou, N., Wang, X., Jia, X., Guo, D., & Xu, B. (2023). Aerogels for Thermal Protection and Their Application in Aerospace. *Gels (Basel, Switzerland)*, *9*(8), 606. DOI: 10.3390/gels9080606 PMID: 37623061

Kausar, A., Ahmad, I., Zhao, T., Aldaghri, O., Ibnaouf, K. H., & Eisa, M. H. (2023). GrapheneNanocomposites as Innovative Materials for Energy Storage and Conversion-Design and Headways. *International Journal of Molecular Sciences*, *24*(14), 11593. DOI: 10.3390/ijms241411593 PMID: 37511354

Khan, N. R., Sharmin, T., & Bin Rashid, A. (2023). Exploring the Versatility of Aerogels: Broad Applications in Biomedical Engineering, Astronautics, Energy Storage, Biosensing, and Current Progress. *Heliyon*, *10*(1), e23102. DOI: 10.1016/j.heliyon.2023.e23102 PMID: 38163169

Lamy-Mendes, A., Lopes, D., Girão, A. V., Silva, R. F., Malfait, W. J., & Durães, L. (2023). Carbon Nanostructures-Silica Aerogel Composites for Adsorption of Organic Pollutants. *Toxics*, *11*(3), 232. DOI: 10.3390/toxics11030232 PMID: 36976997

Lamy-Mendes, A., Torres, R. B., Vareda, J. P., Lopes, D., Ferreira, M., Valente, V., Girão, A. V., Valente, A. J. M., & Durães, L. (2019). Amine Modification of Silica Aerogels/Xerogels for Removal of Relevant Environmental Pollutants. *Molecules (Basel, Switzerland), 24*(20), 3701. DOI: 10.3390/molecules24203701 PMID: 31618901

Lázár, I., Čelko, L., & Menelaou, M. (2023). Aerogel-Based Materials in Bone and Cartilage Tissue Engineering-A Review with Future Implications. *Gels (Basel, Switzerland), 9*(9), 746. DOI: 10.3390/gels9090746 PMID: 37754427

Lebedev, A., Suslova, E., Troyankin, A., & Lovskaya, D. (2021). Investigation of Aerogel Production Processes: Solvent Exchange under High Pressure Combined with Supercritical Drying in One Apparatus. *Gels (Basel, Switzerland), 7*(1), 4. DOI: 10.3390/gels7010004 PMID: 33466392

Lee, K. H., Arshad, Z., Dahshan, A., Alshareef, M., Alsulami, Q. A., Bibi, A., Lee, E.-J., Nawaz, M., Zubair, U., & Javid, A. (2023). Porous Aerogel Structures as Promising Materials for Photocatalysis, Thermal Insulation Textiles, and Technical Applications: A Review. *Catalysts, 13*(9), 1286. DOI: 10.3390/catal13091286

Li, L., Xu, T., Zhang, F., Du, C., & He, S. (2023). Preparation of Super-Flexible Silica Aerogel and Its Application in Oil-Water Separation. *Gels (Basel, Switzerland), 9*(9), 739. DOI: 10.3390/gels9090739 PMID: 37754420

Mathew, J. T., Inobeme, A., Abdulkadir, A., Etsuyankpa, M. B., Shaba, E. Y., Musah, M., Azeh, Y., Adetunji, C. O., Ibrahim, A. M., Tanko, E., & Mamman, A. (2024). a. Properties and Characteristics of Nanocoatings for Medicinal Applications. In book: Sustainable Approach to Protective Nanocoatings. IGI Global; publishing Tomorrow's. *Advances in Chemical and Materials Engineering*, •••, 203–227. DOI: 10.4018/979-8-3693-3136-1.ch007

Mathew, J. T., Inobeme, A., Musah, M., Azeh, Y., Abdullahi, A., Shaba, E. Y., Salihu, A. M., Muhammad, E. B., Josiah, J. G., Jibrin, N. A., Ismail, H., Muhammad, A. I., Maurice, J., Mamman, A., & Ndamitso, M. M. (2024). b. A Critical Review of Green Approach on Wastewater Treatment Strategies. *Journal of Applied Science & Environmental Management*, *28*(2), 363–391. DOI: 10.4314/jasem.v28i2.9

Mathew, J. T., Musah, M., Azeh, Y., & Muhammed, M. (2023). a. Kinetic Study of Heavy Metals Removal from Pharmaceutical Wastewater Using Geopolymer/Fe_3O_4Nanocomposite. *Bima Journal of Science and Technology*, *7*(4), 152–163. DOI: 10.56892/bima.v7i4.539

Mathew, J. T., Musah, M., Azeh, Y., & Muhammed, M. (2023). b. Adsorptive Removal of Selected Toxic Metals from Pharmaceutical Wastewater using Fe_3O_4/ZnONanocomposite. *Dutse Journal of Pure and Applied Sciences*, *9*(4a), 236–248. DOI: 10.4314/dujopas.v9i4a.22

Nguyen, P. T. T., Do, N. H. N., Goh, X. Y., Goh, C. J., Ong, R. H., Le, P. K., Phan-Thien, N., & Duong, H. M. (2022). Recent Progresses in Eco-Friendly Fabrication and Applications of Sustainable Aerogels from Various Waste Materials. *Waste and Biomass Valorization*, *13*(4), 1825–1847. DOI: 10.1007/s12649-021-01627-3 PMID: 34745389

Nita, L. E., Ghilan, A., Rusu, A. G., Neamtu, I., & Chiriac, A. P. (2020). New Trends in Bio-Based Aerogels. *Pharmaceutics*, *12*(5), 449. DOI: 10.3390/pharmaceutics12050449 PMID: 32414217

Noman, M. T., Amor, N., Ali, A., Petrik, S., Coufal, R., Adach, K., & Fijalkowski, M. (2021). Aerogels for Biomedical, Energy and Sensing Applications. *Gels (Basel, Switzerland)*, *7*(4), 264. DOI: 10.3390/gels7040264 PMID: 34940324

Priya, A. K., Muruganandam, M., Ali, S. S., & Kornaros, M. (2023). Clean-Up of Heavy Metals from Contaminated Soil by Phytoremediation: A Multidisciplinary and Eco-Friendly Approach. *Toxics*, *11*(5), 422. DOI: 10.3390/toxics11050422 PMID: 37235237

Romero-Montero, A., Valencia-Bermúdez, J. L., Rosas-Meléndez, S. A., Núñez-Tapia, I., Piña-Barba, M. C., Leyva-Gómez, G., & Del Prado-Audelo, M. L. (2023). Biopolymeric Fibrous Aerogels: The Sustainable Alternative for Water Remediation. *Polymers*, *15*(2), 262. DOI: 10.3390/polym15020262 PMID: 36679143

Shah, N., Rehan, T., Li, X., Tetik, H., Yang, G., Zhao, K., & Lin, D. (2021). Magnetic aerogel: An advanced material of high importance. *RSC Advances*, *11*(13), 7187–7204. DOI: 10.1039/D0RA10275J PMID: 35423256

Singh, A., Ahmed, A., Sharma, A., & Arya, S. (2022). Graphene and Its Derivatives: Synthesis and Application in the Electrochemical Detection of Analytes in Sweat. *Biosensors (Basel)*, *12*(10), 910. DOI: 10.3390/bios12100910 PMID: 36291046

Wu, Y., Wang, X., Yao, L., Chang, S., & Wang, X. (2023). Thermal Insulation Mechanism, Preparation, and Modification of Nanocellulose Aerogels: A Review. *Molecules (Basel, Switzerland)*, *28*(15), 5836. DOI: 10.3390/molecules28155836 PMID: 37570806

Zhou, Y., Liu, Y., Zhang, M., Feng, Z., Yu, D. G., & Wang, K. (2022). ElectrospunNanofiber Membranes for Air Filtration: A Review. *Nanomaterials (Basel, Switzerland)*, *12*(7), 1077. DOI: 10.3390/nano12071077 PMID: 35407195

Chapter 2
Aerogel Nanocomposites: Properties and Applications

John Tsado Mathew
https://orcid.org/0000-0001-9514-5330
Department of Chemistry, Ibrahimm Badamasi Babangida Lapai, Nigeria

Elijah Yanda Shaba
Department of Chemistry, Federal University of Technology, Minna, Nigeria

Monday Musah
https://orcid.org/0000-0002-3224-1664
Department of Chemistry, Ibrahim Badamasi Babangida University, Lapai, Nigeria

Yakubu Azeh
https://orcid.org/0000-0002-3818-674X
Department of Chemistry, Ibrahim Badamasi Babangida University, Lapai, Nigeria

Etsuyankpa Bini Muhammad
Department of Chemistry, Federal University, Lafia, Nigeria

Ezekiel Tanko
Department of Chemistry, Federal University of Technology, Minna, Nigeria

Musa Safiyanu Tanko
Department of Chemistry, Federal University, Lafia, Nigeria

Aishetu Ibrahim Muhammad
Department of Chemistry, Ibrahim Badamasi Babangida University, Lapai, Nigeria

Jonathan Hussaini
https://orcid.org/0009-0004-9216-8691
Department of Chemistry, Ibrahim Badamasi Babangida University, Lapai, Nigeria

DOI: 10.4018/979-8-3693-5971-6.ch002

Amos Mamman

Department of Chemistry, Ibrahim Badamasi Babangida University, Lapai, Nigeria

ABSTRACT

Aerogel nanocomposites, characterized by their porous and lightweight structure, exhibit a diverse array of properties and find applications across various fields. By integrating aerogels with nanomaterials, such as carbon nanotubes, graphene, or metal nanoparticles, these composites achieve enhanced mechanical strength, thermal conductivity, and electrical conductivity compared to traditional aerogels. They often display superior thermal and chemical stability, making them suitable for extreme environments. In terms of applications, aerogel nanocomposites are utilized in insulation for buildings, aerospace materials, energy storage devices, and environmental remediation technologies. Their high surface area and porosity enable efficient adsorption of pollutants from air and water. Furthermore, their lightweight nature makes them attractive for lightweight structural components in vehicles and electronics. Continued research in the synthesis methods and characterization techniques of aerogel nanocomposites is essential to unlock their full potential and expand their applications further

10.1 INTRODUCTION

In recent years, aerogel nanocomposites have attracted significant attention due to their unique properties and versatile applications across various fields. Aerogels are renowned for their ultralow density and high porosity, which contribute to their outstanding thermal insulation capabilities. These attributes have made them valuable in applications requiring superior insulating performance. Despite these benefits, aerogels have traditionally been limited by their brittleness and poor mechanical strength, which restrict their practical use in many applications where durability and load-bearing capacity are essential. The incorporation of nanomaterials into aerogel matrices presents a promising approach to addressing these limitations. By integrating nanomaterials such as carbon nanotubes, graphene, or metal nanoparticles, researchers can significantly enhance the mechanical strength and flexibility of aerogels. This modification improves their resilience and enables them to withstand greater mechanical stress, expanding their potential applications. Additionally, nanomaterials can introduce new functionalities, such as improved thermal and electrical conductivity, or catalytic properties. This advancement opens up new possibilities for aerogel nanocomposites in fields ranging from construction and aerospace to energy storage and environmental remediation. The ongoing development of these materials aims to combine the best of aerogel's intrinsic properties with the added benefits of nanotechnology, broadening their scope of use and enhancing their practical applicability (Bakhori et al. 2023).

Aerogels are exceptional materials characterized by their highly porous structure, which results from the gelation of a precursor solution followed by a supercritical drying process. This method creates a network of interconnected nanostructured pores, endowing aerogels with a distinctive morphology. The resulting structure imparts aerogels with several remarkable properties, such as ultralow density, high surface area, and outstanding thermal insulation capabilities. These attributes make aerogels particularly effective for applications requiring excellent insulating properties and significant surface area for interaction. Despite these advantages, aerogels are inherently fragile due to their delicate, porous network. This fragility limits their use in applications that require structural load-bearing capabilities. Their brittleness can make them prone to damage under mechanical stress or impact, which poses challenges for their incorporation into load-bearing structures. Consequently, while aerogels offer exceptional performance in areas such as insulation and lightweight materials, their practical use in applications requiring mechanical strength and durability remains constrained. Ongoing research aims to address these limitations by enhancing the mechanical robustness of aerogels, thereby expanding their potential applications in more demanding environments and industries (Meti et al. 2023).

Nanomaterials, including carbon nanotubes, graphene, metal nanoparticles, and specialized polymers, provide a valuable approach to reinforcing aerogel structures and adding diverse functionalities. By incorporating these nanomaterials into the aerogel matrix, it is possible to significantly enhance the material's properties to meet specific application needs. For instance, carbon nanotubes and graphene, with their exceptional strength and high aspect ratio, can greatly improve the mechanical strength and flexibility of aerogels. Their strong interfacial interactions with the aerogel matrix contribute to increased structural integrity and durability. Metal nanoparticles, known for their enhanced thermal and electrical conductivity, can enhance heat transfer and electrical performance within aerogels. This makes them suitable for applications requiring efficient thermal management or electrical conductivity. Polymers added to the aerogel matrix can further improve durability and stability by enhancing the material's resistance to environmental factors and mechanical stress. This combination of nanomaterials allows for the tailoring of aerogel nanocomposites to exhibit superior mechanical properties, thermal and electrical conductivity, and overall stability. Consequently, these enhanced aerogels find applications in a variety of fields, including advanced materials, electronics, and environmental technologies, where specialized performance characteristics are crucial (Slepičková et al. 2021).

Incorporating nanomaterials into aerogels imparts a range of advantageous properties that enhance their performance across various applications. For example, carbon-based nanomaterials such as carbon nanotubes and graphene are known for their remarkable mechanical properties. Their high aspect ratio and robust interfacial interactions with the aerogel matrix significantly bolster the material's strength and durability, making them ideal for applications requiring structural integrity and resilience. Metal nanoparticles, on the other hand, contribute to improved thermal conductivity. Their presence enhances phonon scattering within the aerogel, facilitating better heat transfer and thermal management. This property is particularly beneficial in applications where efficient heat dissipation is critical, such as in thermal protection systems or electronic devices. Additionally, metal nanoparticles can endow aerogels with catalytic functionality, making them valuable for energy conversion processes and environmental remediation. These nanoparticles can act as catalysts in chemical reactions, improving the efficiency of processes such as hydrogen production or pollutant degradation. By integrating these nanomaterials, aerogels can be tailored to meet specific performance requirements, broadening their applicability in diverse fields such as construction, aerospace, energy, and environmental technology (Speranza, 2021; Inobeme et al., 2024a).

The unique properties of aerogel nanocomposites make them highly versatile, suitable for a wide range of applications. In the construction industry, aerogel-based insulation materials are particularly valuable due to their exceptional thermal performance. These materials significantly enhance energy efficiency by reducing heat transfer, which leads to lower energy consumption and improved comfort within buildings. Their superior insulating properties help in maintaining consistent indoor temperatures and minimizing heating and cooling costs. In aerospace applications, aerogel nanocomposites are utilized for their lightweight and insulating capabilities. They are employed in thermal protection systems to safeguard spacecraft and aircraft from extreme temperatures, as well as in cryogenic insulation to maintain low temperatures in space missions. Their lightweight nature also makes them ideal for structural components, where reducing weight is crucial for enhancing performance and fuel efficiency. Additionally, aerogel nanocomposites are increasingly used in energy storage devices such as batteries and supercapacitors. Their high surface area and porous structure facilitate efficient ion transport and storage, which improves the performance and energy density of these devices. By leveraging their distinctive properties, aerogel nanocomposites contribute to advancements in construction, aerospace, and energy storage technologies, highlighting their broad applicability and potential for innovation (Ganobjak et al. 2023).

Aerogel nanocomposites offer significant promise for environmental remediation, especially in the removal of pollutants from air and water. Their unique properties, such as an exceptionally high surface area and a highly porous structure, make them highly effective for adsorbing a wide range of contaminants. This includes volatile organic compounds (VOCs), heavy metals, and substances associated with oil spills. The large surface area of aerogels provides ample space for contaminants to adhere, while the porous nature allows for extensive interaction with the pollutants, enhancing the efficiency of the adsorption process. Moreover, the surface chemistry of aerogels can be finely tuned, enabling the functionalization of nanocomposites to target specific pollutants more effectively. By modifying the surface properties, researchers can design aerogel nanocomposites that selectively capture particular types of contaminants, further improving their efficacy in environmental cleanup efforts. This customization makes aerogel nanocomposites valuable tools in addressing diverse pollution challenges. As such, they hold potential for a range of applications, including air purification, water treatment, and spill response. Continued development and optimization of these materials will enhance their effectiveness and broaden their application in environmental protection and remediation strategies (Asghar et al. 2024).

Despite their considerable potential, the development and widespread adoption of aerogel nanocomposites face several significant challenges. One major hurdle is the scalability of synthesis methods, which must be optimized to produce aerogel nanocomposites efficiently and cost-effectively on a larger scale. Additionally, achieving precise control over the dispersion and alignment of nanomaterials within the aerogel matrix is crucial for ensuring consistent performance and desired properties. Concerns about cost also remain, as high production expenses can limit the commercial viability of these materials. Furthermore, addressing the environmental impact and health and safety considerations associated with nanomaterials is essential. The production and use of nanomaterials can raise environmental and health risks, which must be managed through careful regulation and safety protocols. Overcoming these challenges will require a concerted interdisciplinary approach, integrating expertise from materials science, chemistry, engineering, and environmental science. Collaborative efforts in these fields are necessary to advance the synthesis techniques, improve characterization methods, and develop practical applications for aerogel nanocomposites. By addressing these issues comprehensively, it will be possible to unlock the full potential of aerogel nanocomposites and enable their broader adoption in various industries, driving innovation and solving critical problems in energy, environment, and materials technology (Caddeo et al. 2020; Mathew et al., 2023a).

In general, aerogel nanocomposites represent a highly promising class of materials, thanks to their tailored properties and wide range of applications. These materials offer unique advantages, including low density, high surface area, and customizable characteristics, making them suitable for various advanced applications. Ongoing research and innovation in aerogel nanocomposites have the potential to tackle pressing challenges in critical areas such as energy efficiency, where they can improve insulation and energy storage; environmental sustainability, through applications in pollution control and resource conservation; and advanced materials development, by enabling the creation of novel materials with enhanced performance. The continued exploration of aerogel nanocomposites promises to drive significant advancements and address key issues across multiple sectors (**Figure 1**) (Khan et al. 2023).

Figure 1. Integration of nanomaterials into aerogel

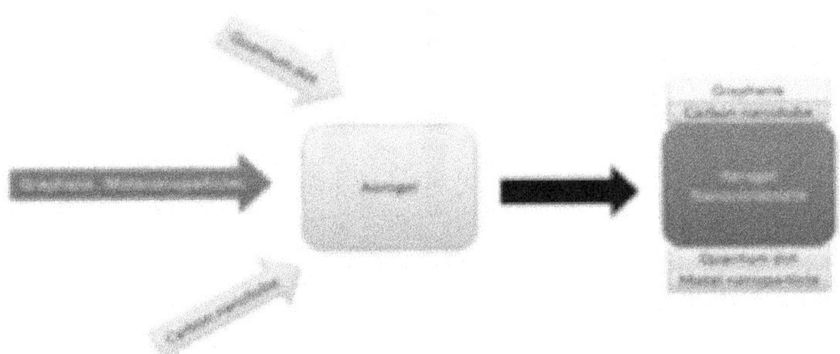

10.2 PROPERTIES OF AEROGEL NANOCOMPOSITES

This table 1 provides a concise overview of the key properties of aerogel nanocomposites, including mechanical strength, thermal conductivity, electrical conductivity, and chemical stability, along with examples of nanomaterials used to achieve these properties and potential applications in various fields. Aerogel nanocomposites exhibit enhanced properties compared to traditional aerogels due to the incorporation of nanomaterials into their structure. One significant improvement is in mechanical strength (**figure 2**), where nanomaterials like carbon nanotubes and graphene reinforce the aerogel matrix, resulting in increased tensile, compressive, and flexural strength. This enhancement makes aerogel nanocomposites suitable for applications requiring structural integrity, such as aerospace components and lightweight structural materials (Owais et al. 2022; Mathew et al., 2023b).

Figure 2. Properties of aerogel nanocomposites

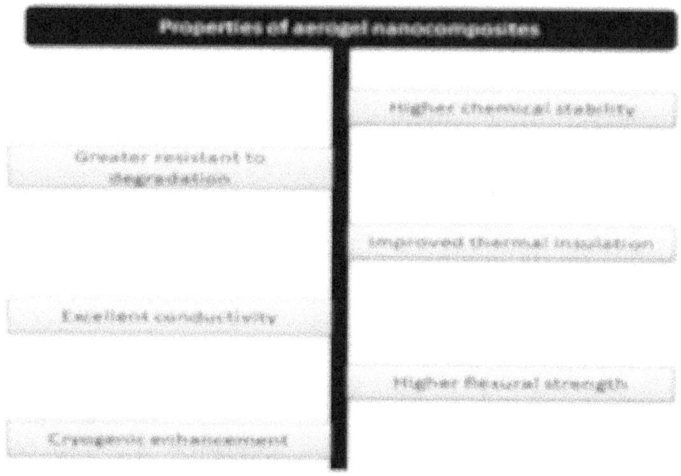

Another notable property is thermal conductivity, which can be improved by dispersing metal nanoparticles or carbon-based nanomaterials throughout the aerogel matrix. This enhancement enables better thermal insulation performance and regulation, making aerogel nanocomposites ideal for use in insulation materials and thermal protection systems. Furthermore, the incorporation of conductive nanomaterials like graphene or conductive polymers enhances the electrical conductivity of aerogel nanocomposites. This property opens up opportunities for applications in electronics, sensors, and flexible electronics where efficient electrical conduction is crucial (Ghica et al. 2023; Inobeme et al. 2024b).

Lastly, aerogel nanocomposites often demonstrate increased chemical stability, which enhances their resistance to degradation and environmental exposure. This improved stability broadens their applicability across various fields, including environmental remediation, where they can be used to remove pollutants; catalysis, where their durability supports long-term catalytic processes; and protective coatings, where they offer enhanced resistance to harsh conditions. The ability of aerogel nanocomposites to withstand chemical and environmental challenges makes them versatile materials with potential for innovative applications in these critical areas (Chen et al. 2023).

Table 1. The properties of aerogel nanocomposites

Property	Description	Example Nanomaterials	Applications	References
Mechanical Strength	Enhanced tensile, compressive, and flexural strength compared to traditional aerogels.	Carbon nanotubes (CNTs), graphene, polymer nanofibers	Aerospace components, lightweight structural materials, protective clothing	Qi et al. 2023.
Thermal Conductivity	Improved thermal insulation and conductivity, facilitating heat transfer regulation.	Metal nanoparticles, carbon-based nanomaterials	Insulation materials, thermal protection systems, cryogenic insulation	Jóźwiak et al. 2020.
Electrical Conductivity	Enhanced electrical conductivity, enabling applications in electronics and sensors.	Graphene, conductive polymers	Flexible electronics, conductive coatings, sensors	Han et al. 2021; Inobeme et al. 2024c
Chemical Stability	Increased resistance to chemical degradation and environmental exposure.	Functionalized nanoparticles, polymer coatings	Environmental remediation, catalysis, protective coatings	Grisolia et al. 2023.

10.3 SYNTHESIS METHODS

This **Table 2** provides a summary of different synthesis methods for aerogel nanocomposites, including the sol-gel process, supercritical drying, nanomaterial incorporation techniques, and template-assisted synthesis. Each method for tailoring material properties presents unique advantages and disadvantages related to control over composition, porosity, processing time, equipment requirements, and pore structure. For instance, some techniques offer precise control over the material's composition and pore size, while others may be more time-efficient but provide less control over these factors. Equipment requirements can vary significantly, influencing the feasibility and cost of implementation. The choice of method depends on the specific needs of the application, such as the desired porosity and structural characteristics, as well as practical considerations like processing speed and available resources (Shah et al. 2021).

Synthesis methods for aerogel nanocomposites include the sol-gel process, supercritical drying, nanomaterial incorporation, and template-assisted synthesis. The sol-gel process involves the hydrolysis and condensation of precursors to form a gel, offering control over composition and porosity. Supercritical drying replaces the liquid phase with a supercritical fluid to preserve the porous structure. Nanomaterial incorporation techniques ensure uniform dispersion of nanoparticles within the aerogel matrix, enhancing properties like mechanical strength and conductivity. Template-assisted synthesis utilizes sacrificial templates to control pore size and morphology during aerogel formation, enabling precise control over pore structure. Each method has its advantages and limitations, ranging from control over properties and structure to equipment requirements and processing time. The choice of synthesis method depends on the desired properties and intended applications of the aerogel nanocomposites (Barrios et al. 2019; Mathew et al., 2024a).

Table 2. Synthesis methods

Synthesis Method	Description	Advantages	Disadvantages	References
Sol-Gel Process	Involves the hydrolysis and condensation of precursor materials to form a gel, followed by drying.	Control over composition and porosity	Long processing time, solvent removal	Kamanina et al. 2022
Supercritical Drying	Utilizes supercritical fluid to replace the liquid phase in the gel, preventing collapse of the structure.	Preserves the porous structure	Requires specialized equipment	Şahin et al. 2017; Inobeme et al. 2024d
Nanomaterial Incorporation	Incorporates nanomaterials into the aerogel matrix during gelation or post-gelation stages.	Uniform dispersion of nanomaterials	Potential agglomeration of nanoparticles	Adeyemi & Fawole, 2023.
Template-Assisted Synthesis	Utilizes sacrificial templates to control pore size and morphology during aerogel formation.	Precise control over pore structure	Template removal can be challenging	Wu et al. 2020.

10.4 CHARACTERIZATION TECHNIQUES

Characterization techniques play a crucial role in understanding the properties and structure of aerogel nanocomposites, aiding in the optimization of synthesis methods and the development of new applications. Here, we will discuss several key characterization techniques commonly employed in the study of aerogel nanocomposites (Long et al. 2018).

Scanning Electron Microscopy (SEM): Scanning Electron Microscopy (SEM) is a high-resolution imaging technique employed to examine the surface morphology and microstructure of aerogel nanocomposites. In SEM, a focused electron beam is directed at the sample, and the emitted secondary electrons are detected to generate detailed images. This imaging method offers magnifications ranging from 10x to 100,000x, allowing researchers to observe fine details with exceptional clarity. SEM is particularly useful for analyzing features such as pore size, particle distribution, and the dispersion of nanomaterials within the aerogel matrix. By providing high-resolution images of the sample's surface, SEM enables a thorough examination of the material's microstructural characteristics. This detailed observation is crucial for understanding the uniformity and arrangement of nanoparticles, as well as for assessing the effectiveness of the aerogel's structural design. SEM's ability to visualize and measure these features helps in optimizing material properties and ensuring the desired performance of aerogel nanocomposites in various applications, including filtration, insulation, and catalysis (Juhász et al. 2021; Olorunsola et al. 2024).

Transmission Electron Microscopy (TEM): Transmission Electron Microscopy (TEM) provides superior resolution compared to Scanning Electron Microscopy (SEM), enabling the visualization of nanoscale structures within aerogel nanocomposites. In TEM, electrons are transmitted through ultrathin sections of the sample, generating high-resolution images with atomic-scale detail. This allows researchers to observe and analyze the size, shape, and crystallinity of nanoparticles within the aerogel matrix. TEM's ability to achieve such fine resolution is crucial for understanding the precise structural characteristics of nanocomposites. Furthermore, TEM can be employed for Selected Area Electron Diffraction (SAED), a technique that provides additional insights into the crystallographic orientations and phases present in the sample. SAED helps in identifying the specific crystal structures and phases of the nanomaterials, offering valuable information about their arrangement and properties. Together, these capabilities make TEM an essential tool for in-depth structural analysis of aerogel nanocomposites, facilitating the optimization of material properties and the development of advanced applications in fields such as nanotechnology, materials science, and electronics (Guzzinati et al. 2018).

Brunauer-Emmett-Teller (BET) Analysis: Brunauer-Emmett-Teller (BET) analysis is a widely utilized technique for determining the specific surface area and pore size distribution of aerogel nanocomposites. This method is based on the adsorption of gas molecules onto the surface of the sample at various relative pressures. During the analysis, the sample is exposed to a gas, typically nitrogen, and the amount of gas adsorbed is measured as a function of the pressure. By analyzing these adsorption data, BET analysis calculates key parameters such as the specific surface area and pore volume of the material. This provides detailed insights into the porosity and internal structure of the aerogel nanocomposite. The specific surface area reflects the extent of the material's available surface for interactions, while the pore size distribution reveals the range and distribution of pore sizes within the material. These characteristics are critical for applications where surface area and pore structure play a significant role, such as in catalysis, filtration, and energy storage. BET analysis thus offers essential information for optimizing material performance and tailoring aerogel nanocomposites for specific functional applications (Horvat et al. 2022).

Fourier Transform Infrared (FTIR) Spectroscopy: Fourier Transform Infrared (FTIR) spectroscopy is a powerful technique used to analyze the chemical composition and identify functional groups in aerogel nanocomposites. This method operates by measuring the absorption of infrared radiation, which interacts with the molecular vibrations within the sample. As infrared light passes through the aerogel, specific wavelengths are absorbed at frequencies corresponding to the vibrational modes of different chemical bonds. By examining the intensity and frequency of these absorption bands in the infrared spectrum, FTIR spectroscopy can identify characteristic bonds and functional groups present in the material. This analysis allows researchers to determine the types of nanomaterials incorporated into the aerogel, as well as to detect any surface modifications or interactions within the aerogel matrix. Such insights are crucial for understanding the chemical interactions and properties of the nanocomposite, which can influence its performance in various applications. FTIR spectroscopy thus provides valuable information for optimizing material formulations, ensuring compatibility of components, and enhancing the overall functionality of aerogel nanocomposites in diverse technological contexts (Barnes et al. 2023).

X-ray Diffraction (XRD): X-ray Diffraction (XRD) is a fundamental technique used to analyze the crystalline structure and phase composition of aerogel nanocomposites. In an XRD analysis, the sample is exposed to monochromatic X-ray radiation, which interacts with the material's crystal lattice. This interaction results in the scattering of X-rays, producing a diffraction pattern that reflects the atomic arrangement within the material. By examining these diffraction patterns, researchers can identify the crystalline phases present in the nanocomposite and determine key crystallographic parameters, such as lattice spacing and unit cell dimensions. Additionally, XRD provides insights into the degree of crystallinity, which indicates the proportion of ordered versus disordered regions within the material. These data are essential for understanding the structural properties of the nanocomposite, which influence its mechanical, thermal, and chemical behavior. Through XRD, researchers can optimize material properties and ensure that the nanocomposite meets specific performance requirements for its intended applications. This comprehensive structural analysis is crucial for advancing the development and application of aerogel nanocomposites in various technological fields (Maponya et al. 2022).

Thermogravimetric Analysis (TGA): Thermogravimetric Analysis (TGA) is a pivotal thermal analysis technique used to investigate the thermal stability and decomposition behavior of aerogel nanocomposites. During a TGA test, the sample is subjected to controlled temperature ramps while being exposed to an inert atmosphere. This procedure measures changes in the sample's mass as a function of temperature, offering critical insights into its thermal properties. Key parameters determined through TGA include the onset temperature of decomposition, which indicates the temperature at which the material begins to break down. The weight loss kinetics reveal how the material's mass changes with increasing temperature, providing information on the rate and extent of decomposition. Additionally, the residual mass after thermal treatment shows the amount of material that remains intact, helping researchers understand the material's stability and composition. By analyzing these factors, TGA provides valuable data on the thermal behavior and durability of aerogel nanocomposites, which is essential for applications requiring high-temperature performance and stability. This information is crucial for optimizing material properties and ensuring their effectiveness in demanding environments (Azeh et al., 2023).

Dynamic Mechanical Analysis (DMA): Dynamic Mechanical Analysis (DMA) is a critical technique used to evaluate the mechanical properties and viscoelastic behavior of aerogel nanocomposites across different temperatures and frequencies. During a DMA test, the sample is subjected to sinusoidal mechanical stress or strain, and the resulting mechanical response is simultaneously measured. This process provides valuable data on several key parameters, including the storage modulus, loss modulus, and damping behavior. The storage modulus reflects the material's stiffness, indicating how well it can maintain its shape under stress. The loss modulus represents the energy dissipated as heat, which is crucial for understanding the material's internal friction and energy absorption characteristics. Damping behavior is assessed to determine how effectively the material can absorb and dissipate mechanical vibrations. By analyzing these parameters, researchers can gain insights into the stiffness, resilience, and damping properties of the nanocomposite. This information is essential for optimizing the design of materials tailored for specific mechanical applications, ensuring they meet performance requirements and function effectively in their intended environments (Cristea et al. 2020; Inobeme et al., 2023).

Furthermore, employing a combination of characterization techniques enables researchers to thoroughly analyze the properties, structure, and performance of aerogel nanocomposites. Techniques such as spectroscopy, microscopy, and mechanical testing provide a detailed understanding of the material's composition, morphology, and functional attributes. By integrating these insights, researchers can refine synthesis methods, optimize material properties, and explore innovative applications for aerogel nanocomposites. For instance, enhanced characterization can lead to improvements in aerogel performance for applications in insulation, where excellent thermal resistance is crucial. In aerospace, precise material understanding contributes to the development of lightweight, high-strength components. In energy storage, characterization helps in designing aerogels with optimal porosity and surface area for efficient energy absorption and release. Additionally, environmental remediation applications benefit from tailored properties that enhance the material's ability to absorb pollutants. Overall, leveraging a comprehensive suite of characterization techniques allows for the advancement of aerogel nanocomposites, leading to breakthroughs in various fields and driving the development of next-generation materials that address critical technological and environmental challenges (Lázár & Menelaou, 2023).

10.5 APPLICATIONS OF AEROGEL NANOCOMPOSITES

The **Table 3** below provides a summary of various applications of aerogel nanocomposites, including insulation materials, aerospace components, energy storage devices, and environmental remediation. Each application benefits from the unique properties of aerogel nanocomposites, such as high thermal resistance, lightweight, mechanical strength, and environmental stability, enabling their use in diverse fields for improving performance and efficiency (Khan et al. 2023).

Aerogel nanocomposites are valued across various industries due to their unique and versatile properties. In the realm of insulation, they provide exceptional thermal resistance, which helps reduce energy consumption in both residential and industrial settings. Their low thermal conductivity makes them ideal for insulating buildings, pipelines, and equipment, leading to significant energy savings. In aerospace, aerogel nanocomposites offer lightweight thermal protection and structural components for spacecraft and aircraft, where reducing weight while maintaining thermal efficiency is critical for performance and safety. Energy storage devices also benefit from their high surface area and enhanced conductivity, which improve the efficiency and capacity of batteries and supercapacitors. The high porosity and adsorption capacity of aerogels make them effective in environmental remediation applications, where they are used to remove pollutants from air and water. By leveraging these unique properties, aerogel nanocomposites address a range of technological and environmental challenges, offering innovative solutions in insulation, aerospace engineering, energy storage, and pollution control. Their adaptability and performance continue to drive advancements and new applications in these fields (Noman et al. 2021; Mathew et al. 2024b).

Their applications extend to protective clothing, where their lightweight yet robust nature offers enhanced safety for workers in hazardous environments. Additionally, aerogel nanocomposites are used in electronics for their electrical conductivity and insulation properties, enabling the development of efficient and compact electronic devices. In the automotive industry, they contribute to lightweighting efforts, improving fuel efficiency and reducing emissions. Overall, aerogel nanocomposites play a crucial role in various sectors, offering solutions to challenges related to energy efficiency, environmental sustainability, safety, and performance. Continued research and development in this field hold the promise of unlocking further applications and advancing technological innovation (Bhattacharjee et al. 2019).

Table 3. The outlining various applications of aerogel nanocomposites

Application	Description	Example Nanomaterials	Advantages	References
Insulation Materials	Provide superior thermal insulation for buildings, appliances, and industrial equipment, reducing energy consumption and improving comfort.	Carbon nanotubes, graphene, metal nanoparticles	High thermal resistance, lightweight, thin profiles for space-constrained applications	AlMarri et al. 2021.
Aerospace Components	Utilized in thermal protection systems, cryogenic insulation, and lightweight structural components for spacecraft and aircraft.	Silica aerogels, carbon nanotubes, graphene	High strength-to-weight ratio, thermal stability, resistance to extreme environments	Borella et al. 2023.
Energy Storage Devices	Employed in batteries, supercapacitors, and fuel cells to enhance energy storage capacity, efficiency, and cycling stability.	Conductive polymers, metal oxides, carbon-based nanomaterials	High surface area for improved ion transport, enhanced electrical conductivity, prolonged cycle life	AnilKumar et al. 2023.
Environmental Remediation	Used for the adsorption and removal of pollutants from air and water, including volatile organic compounds (VOCs) and heavy metals.	Functionalized nanoparticles, carbon-based nanomaterials	High surface area and porosity for efficient adsorption, tunable surface chemistry for selective pollutant capture	Guerra et al. 2018; Mathew et al., 2024c

10.6 TAILORING PROPERTIES FOR SPECIFIC APPLICATIONS

In various industries ranging from electronics to construction, tailoring material properties for specific applications is paramount for achieving desired performance and functionality. This process involves understanding the requirements of the application and then modifying material characteristics such as strength, conductivity, and durability to meet those needs. Whether it's enhancing thermal properties for aerospace components or optimizing electrical conductivity for semiconductor devices, the ability to tailor material properties is a fundamental aspect of modern engineering. In this article, we'll explore the importance of tailoring properties for specific applications and delve into some common methods used to achieve this (Alshangiti et al. 2023).

One of the key motivations for tailoring material properties is to enhance performance while reducing costs and conserving resources. In industries such as automotive manufacturing, the demand for lightweight yet strong materials are driven by the need to improve fuel efficiency without sacrificing safety. By adjusting the properties of materials like aluminum alloys or carbon fiber composites, engineers can create components that meet the strict requirements for vehicle structures while significantly reducing the overall weight. This weight reduction not only improves the vehicle's handling and performance but also contributes to sustainability initiatives by lowering fuel consumption and greenhouse gas emissions. Lighter vehicles consume less fuel, which in turn leads to fewer emissions and a reduced carbon footprint. Additionally, optimizing material usage aligns with broader efforts to minimize waste and conserve natural resources, further supporting long-term sustainability goals. By carefully balancing material properties, engineers can address both performance demands and environmental concerns, ultimately advancing the development of more efficient and eco-friendly technologies (Sayam et al. 2022).

In aerospace applications, materials are subjected to extreme conditions such as high temperatures, intense pressure differentials, and rapid thermal cycling. As a result, materials must demonstrate exceptional thermal stability and mechanical strength to perform reliably under these harsh environments. Tailoring material properties through methods like alloying, heat treatment, and the application of specialized coatings enables engineers to create materials capable of enduring the severe demands of space travel and high-speed flight. For instance, nickel-based superalloys are commonly used in turbine engines because of their remarkable strength and resistance to thermal degradation at elevated temperatures. These properties are achieved through the careful control of the alloy's composition and the microstructure, which is refined through precise manufacturing techniques. Additionally, advanced coatings can be applied to further enhance the material's resistance to oxidation, corrosion, and wear, extending the lifespan of aerospace components. By optimizing material properties in this way, engineers ensure that critical aerospace systems maintain their integrity and functionality even in the most challenging operating conditions (Suryanarayana, 2019; Mathew et al. 2024d).

In the rapidly advancing field of electronics, where miniaturization and performance are paramount, tailoring material properties is crucial for the development of cutting-edge semiconductor devices. Semiconductor materials, such as silicon, are meticulously engineered to exhibit specific electrical conductivity and bandgap properties, enabling precise control over electron flow—an essential requirement for the efficient functioning of transistors and integrated circuits. Techniques such as doping, which introduces impurities to modify electrical characteristics, and epitaxy, which creates thin crystalline layers, allow manufacturers to fine-tune material properties at the nanoscale. This precise control at such a small scale is critical for the fabrication of ever-smaller, more powerful, and energy-efficient electronic devices. As electronics continue to shrink in size while increasing in complexity and functionality, the ability to manipulate material properties becomes even more important. These advancements enable the creation of high-performance devices like microprocessors, memory chips, and sensors that power modern technology, from smartphones to sophisticated computing systems, all while improving energy efficiency and extending battery life. (Filipovic & Grasser, 2022).

In the construction industry, the growing demand for sustainable building materials has driven significant innovation in tailoring material properties to enhance both durability and energy efficiency. For example, concrete, one of the most widely used construction materials, can be modified by incorporating supplementary cementitious materials like fly ash or slag, as well as reinforcing fibers, to improve its strength, reduce carbon emissions, and increase resistance to cracking and weathering. These modifications not only enhance the material's performance but also contribute to environmental sustainability by reducing the carbon footprint associated with traditional cement production. Engineers can further customize material formulations based on specific project factors such as local climate conditions, load-bearing requirements, and the desired environmental impact, leading to more efficient and resilient structures. This level of customization ensures that materials are optimized for both functionality and sustainability in modern infrastructure projects. By focusing on these tailored solutions, the construction industry is advancing toward greener practices, creating buildings and infrastructure that are not only longer-lasting but also more environmentally responsible (Shah & Huseien, 2020).

Various methods are used to tailor material properties for specific applications, spanning from traditional metallurgical techniques to cutting-edge nanotechnology. These methods include alloying, heat treatment, surface modification, nanomaterial synthesis, and composite fabrication, each offering distinct advantages and challenges based on the target properties and application needs. Alloying, for example, allows for precise control of a material's composition, enabling engineers to achieve desired mechanical, thermal, and chemical properties suited for diverse applications. Heat treatment can further enhance properties such as hardness, strength, and ductility by manipulating a material's microstructure. Meanwhile, surface modification techniques, such as coating or plasma treatment, can improve a material's resistance to wear, corrosion, or friction. On the more advanced side, nanotechnology enables the manipulation of material behavior at the atomic and molecular levels, offering unparalleled precision in tailoring properties. This can lead to breakthroughs in fields like electronics, medicine, and energy storage, where ultra-small-scale properties are crucial. Together, these approaches enable the development of materials that are optimized for a wide range of innovative applications, pushing the boundaries of performance and functionality (Ogbezode et al. 2023).

Tailoring material properties for specific applications is a cornerstone of modern engineering, playing a crucial role across various industries. Whether it's optimizing strength and thermal stability for aerospace components, enhancing electrical conductivity and miniaturization for semiconductor devices, or improving energy efficiency and durability in construction materials, the ability to customize material characteristics allows engineers to develop innovative solutions to complex challenges. This customization enables materials to meet stringent performance requirements while also addressing factors such as sustainability, cost, and resource efficiency. As technology continues to evolve, the demand for advanced, tailored materials will only increase, driving further research and innovation in materials science and engineering. By employing a diverse array of techniques—ranging from traditional methods like alloying and heat treatment to cutting-edge approaches in nanotechnology and surface engineering—engineers can push the boundaries of what's possible. These innovations will pave the way for the next generation of high-performance materials, creating solutions that not only meet but exceed the demands of tomorrow's applications in fields like aerospace, electronics, and sustainable construction (Okonta, 2023).

10.7 ENVIRONMENTAL AND HEALTH CONSIDERATIONS

As we delve deeper into tailoring material properties for specific applications, it is essential to consider not only performance and functionality but also the environmental and health implications of these modifications. While the pursuit of advanced materials drives innovation across industries, it must be balanced with a commitment to minimizing negative impacts on ecosystems and human well-being. Tailoring material properties often involves the use of chemicals, energy-intensive processes, and novel substances, which can have unintended environmental consequences, such as pollution, resource depletion, or hazardous waste generation. Similarly, the health risks associated with certain materials, particularly nanomaterials or toxic additives, must be carefully evaluated to prevent harm to workers, consumers, and the environment. In this article, we will examine the environmental and health considerations related to tailoring material properties and explore strategies for mitigating potential risks. These strategies include adopting greener manufacturing techniques, utilizing sustainable and non-toxic materials, and implementing robust safety standards throughout the product lifecycle. By addressing these factors, engineers and scientists can continue to innovate while ensuring that new materials contribute to a healthier, more sustainable future (Parapouli et al. 2020; Adetuyi et al. 2024).

One of the primary environmental concerns related to material modification is the extraction and processing of raw materials, which can have significant ecological and health impacts. Many industrial processes involved in tailoring material properties require substantial energy consumption and often result in the emission of pollutants, such as greenhouse gases, particulate matter, and toxic chemicals. For instance, mining operations for rare earth metals, critical in the production of electronics and renewable energy technologies, can cause habitat destruction, soil and water contamination, and pose serious health risks to nearby communities due to the release of harmful substances. In addition, the production of certain polymers and composite materials frequently involves the use of hazardous chemicals and the release of volatile organic compounds (VOCs) into the atmosphere, contributing to air pollution and posing risks to both human health and the environment. These environmental and health concerns highlight the need for more sustainable approaches to material modification, including the development of cleaner production technologies, the use of renewable or less harmful raw materials, and stricter regulations to minimize pollution and protect ecosystems (Uddin et al. 2022).

To address these environmental challenges, engineers and researchers are increasingly turning to sustainable material alternatives and innovative manufacturing processes. Key strategies include the recycling and reuse of materials, which help to reduce the demand for virgin resources and decrease waste. Cleaner production methods, such as green chemistry, aim to reduce the use of hazardous substances and lower pollution levels, while additive manufacturing (3D printing) enables precise material usage and minimizes excess waste. By integrating principles of the circular economy—where materials are continuously cycled through use, reuse, and recycling—and conducting comprehensive lifecycle assessments, it is possible to evaluate and enhance the environmental performance of materials throughout their entire lifespan. These approaches collectively work to minimize resource depletion, reduce waste generation, and mitigate negative environmental impacts, ensuring that material innovations contribute positively to sustainability goals. This holistic approach not only addresses immediate environmental concerns but also supports long-term ecological balance and resource efficiency (Tarazona et al.2022).

Another crucial consideration is the potential health hazards associated with modified materials, which can arise both during their production and throughout their lifecycle. The use of certain chemical additives and processing agents in material modification may pose significant risks to workers' health and safety if not managed with care. For example, exposure to airborne nanoparticles during the synthesis or handling of nanomaterials can lead to respiratory issues and other adverse health effects. Similarly, the release of hazardous substances from materials into the environment or consumer products presents additional health risks. Proper management practices, such as implementing stringent safety protocols, providing adequate protective equipment, and conducting thorough risk assessments, are essential to mitigate these health hazards. Additionally, continuous monitoring and regulation of hazardous materials are crucial to prevent and address potential health impacts. By addressing these concerns proactively, it is possible to ensure that material modifications not only enhance performance but also safeguard human health throughout the entire lifecycle of the materials (Xuan et al. 2023).

To address health concerns associated with modified materials, stringent regulations and guidelines are established to manage the use and disposal of hazardous substances across various industries. Occupational health and safety measures are critical in minimizing risks to workers involved in material production and processing. These measures include engineering controls, such as ventilation systems and containment solutions, personal protective equipment (PPE) like gloves and respirators, and regular exposure monitoring to ensure safety standards are met. Additionally, comprehensive product labeling and Material Safety Data Sheets (MSDS) are vital for providing detailed information about potential hazards and safe handling procedures. Consumer awareness campaigns further educate the public on the risks associated with modified materials and promote proper usage and disposal practices. By implementing robust regulatory frameworks and promoting informed practices, industries can better protect both workers and consumers from potential health hazards related to material modifications (Adetunji et al. 2024).

Furthermore, incorporating inherent safety features into material design can significantly reduce risks to both human health and the environment. Prioritizing the use of materials with low toxicity and biodegradability helps mitigate potential hazards throughout their lifecycle. By selecting non-toxic or less hazardous alternatives and minimizing the use of harmful chemicals and additives, engineers can develop materials that are safer to manufacture, use, and dispose of. This approach not only enhances safety but also supports environmental sustainability. Additionally, integrating principles of green chemistry into material development processes plays a crucial role in creating eco-friendly alternatives. Green chemistry focuses on designing processes and products that reduce or eliminate the generation of hazardous substances, leading to materials with lower environmental and health impacts. Sustainable design practices, such as selecting renewable resources and promoting recyclability, further contribute to this goal. Together, these strategies ensure that materials are not only effective and functional but also aligned with broader objectives of safety and environmental stewardship. This holistic approach to material design helps create solutions that are both innovative and responsible, addressing the needs of modern applications while safeguarding health and the planet (Bouchaut et al. 2021).

In general, as we continue to innovate and tailor material properties for specific applications, it's imperative to consider the environmental and health implications of these modifications. By adopting sustainable practices, minimizing the use of hazardous substances, and prioritizing the safety of both workers and end-users, we can ensure that the pursuit of advanced materials is aligned with broader goals of environmental protection and human well-being. Through collaboration across disciplines and adherence to best practices, we can create a future where materials innovation goes hand in hand with environmental stewardship and public health (Bontempi et al. 2021).

10.8 ADVANCEMENTS IN NANOMATERIAL INTEGRATION

The integration of nanomaterials into aerogel matrices has markedly advanced, addressing many of the traditional limitations associated with aerogels and enabling the development of new functionalities. This progress has been pivotal in enhancing the performance and expanding the application range of aerogel nanocomposites. Key advancements in this field can be categorized into several crucial areas. Firstly, improved synthesis techniques have been developed, allowing for better dispersion and alignment of nanomaterials within the aerogel matrix. These techniques, including ultrasonic dispersion and advanced surface modification methods, have overcome issues related to nanomaterial agglomeration and enhanced the uniformity of the composite. Secondly, tailored functionalization of nanomaterials has been a game-changer, enabling the customization of aerogel properties to suit specific needs. By modifying the surface chemistry of nanomaterials, researchers can enhance compatibility and introduce desired functionalities, such as increased catalytic activity or improved mechanical reinforcement. Additionally, the exploration of new nanomaterial types, such as quantum dots and 2D materials, has broadened the range of achievable properties and applications. These novel materials contribute unique characteristics, including enhanced optical properties and tunable electronic features. Together, these advancements have addressed many of the inherent limitations of traditional aerogels, paving the way for more versatile and high-performance aerogel nanocomposites suitable for a wide array of applications (Niculescu et al. 2024).

Recent advancements in synthesis techniques have been crucial for the effective integration of nanomaterials into aerogel matrices. Traditional methods such as sol-gel processing and supercritical drying have been refined to allow for better dispersion and alignment of nanomaterials within the aerogel structure. For instance, the introduction of ultrasonic dispersion and high-energy mixing methods has improved the uniform distribution of nanoparticles, preventing agglomeration and ensuring a more homogenous material. Additionally, advancements in surface modification techniques, such as chemical functionalization and plasma treatment, have enhanced the interaction between nanomaterials and the aerogel matrix. These improved synthesis techniques result in aerogel nanocomposites with superior mechanical strength, thermal conductivity, and stability (Barrios et al. 2019).

The ability to tailor the surface chemistry of nanomaterials has opened new avenues for their integration into aerogels. Functionalization involves modifying the surface properties of nanomaterials to enhance their compatibility with the aerogel matrix and to impart specific functionalities. For example, coating metal nanoparticles with organic ligands can improve their dispersion within the aerogel and enhance their catalytic activity. Similarly, functionalizing carbon nanotubes or graphene with chemical groups that bond effectively with the aerogel matrix can significantly improve mechanical reinforcement and electrical conductivity. This tailored approach enables the creation of aerogel nanocomposites with customized properties, making them suitable for a wide range of specialized applications (Ahmad et al. 2022).

The exploration of novel nanomaterials has significantly expanded the capabilities of aerogel nanocomposites, allowing for a broader range of functionalities and applications. While traditional materials such as carbon nanotubes and metal nanoparticles have already proven their value, recent research has introduced exciting new types of nanomaterials that enhance aerogel properties in innovative ways. Quantum dots are one such advancement, providing unique optical properties that are particularly beneficial for applications in sensing and imaging. Their size-tunable light emission and high photostability make them ideal for high-resolution imaging and detecting trace amounts of substances, thus offering valuable tools for both environmental monitoring and medical diagnostics. Transition metal dichalcogenides (TMDs) represent another significant development. These 2D materials are characterized by their high surface area and versatile electronic properties, which can be tuned for specific applications. TMDs enhance the performance of aerogel nanocomposites in electronic devices and energy storage systems, owing to their excellent conductivity and catalytic activity. Biodegradable nanoparticles have also emerged as a key area of interest. By incorporating these environmentally friendly materials into aerogels, researchers are addressing concerns related to the lifecycle and disposal of materials, thus promoting sustainability and reducing environmental impact. This innovation supports the development of greener aerogel nanocomposites, aligning with broader environmental goals (Paras et al. 2023).

One of the most notable advancements in aerogel nanocomposites is the significant enhancement of their mechanical properties. Historically, aerogels were constrained by their inherent brittleness and low mechanical strength, limiting their practical applications. However, the incorporation of advanced nanomaterials, such as carbon nanotubes and graphene, has markedly improved these characteristics. Carbon nanotubes, renowned for their exceptional tensile strength and flexibility, play a crucial role in reinforcing the aerogel matrix. By integrating these nanotubes, the aerogel's structural integrity is greatly enhanced, making it more resilient to mechanical stress and reducing its susceptibility to cracking and deformation. This results in aerogel nanocomposites with improved durability and load-bearing capabilities. Graphene, another breakthrough material, further advances the mechanical properties of aerogels. With its remarkable tensile strength and high thermal conductivity, graphene not only strengthens the aerogel matrix but also enhances its thermal and electrical performance. The combination of graphene's properties with aerogels leads to composites that can endure more demanding operational conditions and offer superior performance across various applications. These advancements in mechanical reinforcement make aerogel nanocomposites suitable for a broader range of applications, including those requiring robust, load-bearing materials capable of withstanding harsh environments (Noman et al. 2021).

The integration of nanomaterials has significantly increased the functional versatility of aerogel nanocomposites. By incorporating nanoparticles with specific properties, aerogels can be engineered to exhibit a range of functionalities. For example, aerogel nanocomposites embedded with magnetic nanoparticles can be used in magnetic separation applications or as responsive materials in magnetic fields. Aerogels doped with photonic nanomaterials can be utilized in optical sensing and imaging due to their unique light interaction properties. This functional versatility makes aerogel nanocomposites highly adaptable to various fields, including environmental monitoring, medical diagnostics, and advanced manufacturing (Khan et al. 2023).

As the integration of nanomaterials into aerogels advances, addressing environmental and safety concerns becomes increasingly important. The development of greener synthesis methods and safer handling practices is critical to ensuring the sustainable use of aerogel nanocomposites. Researchers are focusing on developing non-toxic, biodegradable nanomaterials and employing environmentally friendly synthesis techniques to reduce the ecological footprint of these materials. Additionally, comprehensive safety evaluations are being conducted to assess the potential health risks associated with nanomaterial exposure and to establish guidelines for safe production and use. These efforts aim to balance the benefits of advanced aerogel nanocomposites with responsible environmental and safety practices (Vrabič-Brodnjak, 2023).

10.9 CONCLUSION

The aerogel nanocomposites offer a promising avenue for advancing materials science and engineering across a wide range of applications. Their unique combination of properties, including ultralow density, high surface area, and tunable mechanical, thermal, and electrical properties, makes them attractive for various industries such as aerospace, energy, electronics, and environmental remediation. By incorporating nanomaterials into aerogel matrices, researchers can tailor properties to meet specific application requirements, enabling innovations in lightweight structural materials, high-performance insulation, efficient catalysts, and versatile sensors. As research in this field continues to evolve, further advancements in synthesis techniques, characterization methods, and scalable manufacturing processes will drive the development of next-generation aerogel nanocomposites with even more enhanced properties and broader utility. With ongoing efforts to address challenges related to cost, scalability, and environmental sustainability, aerogel nanocomposites hold tremendous potential to revolutionize diverse sectors and contribute to the advancement of sustainable technologies in the years to come.

REFERENCES

Adetunji, C. O., Mathew, J. T., Inobeme, A., Taiwo, O., Olaniyan, O., Shakira, G., & Wadazani, D. (2024). Application of Biosensors for Detection and Monitoring of Water Quality. In book: Organic and Inorganic Materials Based Sensors, John Wiley & Sons, Inc., 1065-1077, DOI: 10.1371/journal.pone.0247608 PMID: 33770082

Adetuyi, B. O., Mathew, J. T., Inobeme, A., Falana, Y. O., Adetunji, C. O., Shahnawaz, M., Oyewole, O. A., Eniola, K. I. T., Yerima, M. B., & Popoola, O. A. (2024). Tyres, Bitumen Wear and Plastic Bottles, Other Single Used Plastic as Major Sources of Microplastic. In book: Microplastic Pollution. Springer, Singapore, 1, 141–160. DOI: DOI: 10.1007/978-981-99-8357-5_9

Ahmad, F., Salem-Bekhit, M. M., Khan, F., Alshehri, S., Khan, A., Ghoneim, M. M., Wu, H. F., Taha, E. I., & Elbagory, I. (2022). Unique Properties of Surface-Functionalized Nanoparticles for Bio-Application: Functionalization Mechanisms and Importance in Application. *Nanomaterials (Basel, Switzerland)*, *12*(8), 1333. DOI: 10.3390/nano12081333 PMID: 35458041

Al Marri, M. G., Al-Ghouti, M. A., Shunmugasamy, V. C., & Zouari, N. (2021). Date pits based nanomaterials for thermal insulation applications-Towards energy efficient buildings in Qatar. *PLoS One*, *16*(3), e0247608. DOI: 10.1371/journal.pone.0247608 PMID: 33770082

Alshangiti, D. M., El-Damhougy, T. K., Zaher, A., Madani, M., & Mohamady ghobashy, M. (2023). Revolutionizing biomedicine: advancements, applications, and prospects of nanocomposite macromolecular carbohydrate-based hydrogel biomaterials: a review. *RSC Advances*, *13*(50), 35251–35291. DOI: 10.1039/D3RA07391B PMID: 38053691

Anil Kumar, Y., Koyyada, G., Ramachandran, T., Kim, J. H., Sajid, S., Moniruzzaman, M., Alzahmi, S., & Obaidat, I. M. (2023). Carbon Materials as a Conductive Skeleton for Supercapacitor Electrode Applications: A Review. *Nanomaterials (Basel, Switzerland)*, *13*(6), 1049. DOI: 10.3390/nano13061049 PMID: 36985942

Asghar, N., Hussain, A., Nguyen, D. A., Ali, S., Hussain, I., Junejo, A., & Ali, A. (2024). Advancement in nanomaterials for environmental pollutants remediation: A systematic review on bibliometrics analysis, material types, synthesis pathways, and related mechanisms. *Journal of Nanobiotechnology*, *22*(1), 26. DOI: 10.1186/s12951-023-02151-3 PMID: 38200605

Azeh, Y., Musah, M., Mathew, J. T., Badeggi, U. M., Estuyankpa, M. B., Muhammad, Z. D., Oluwaseyi, D. S., Umar, M. T., Abdulkarim, M. A., Muhammad, A. I., Ibrahim, A. A., Ismail, H., Muhammad, K. T., Umar, G., & Ahmed, T. (2023). In Situ Synthesis of Polyaniline Nanohybrid and Formulation of Polyaniline/Carboxymethyl cellulose/Ethylene Glycol Nanocomposite: Study of its Conducting and Antibacterial Properties. *Journal of Materials and Polymer Science*, *3*(4), 1–9. DOI: 10.47485/2832-9384.1041

Bakhori, N. M., Ismail, Z., Hassan, M. Z., & Dolah, R. (2023). Emerging Trends in Nanotechnology: Aerogel-Based Materials for Biomedical Applications. *Nanomaterials (Basel, Switzerland)*, *13*(6), 1063. DOI: 10.3390/nano13061063 PMID: 36985957

Barnes, M., Sulé-Suso, J., Millett, J., & Roach, P. (2023). Fourier transform infrared spectroscopy as a non-destructive method for analysing herbarium specimens. *Biology Letters*, *19*(3), 20220546. DOI: 10.1098/rsbl.2022.0546 PMID: 36946131

Barrios, E., Fox, D., Li Sip, Y. Y., Catarata, R., Calderon, J. E., Azim, N., Afrin, S., Zhang, Z., & Zhai, L. (2019). Nanomaterials in Advanced, High-Performance Aerogel Composites: A Review. *Polymers*, *11*(4), 726. DOI: 10.3390/polym11040726 PMID: 31010008

Bhattacharjee, S., Joshi, R., Chughtai, A. A., & Macintyre, C. R. (2019). Graphene Modified Multifunctional Personal Protective Clothing. *Advanced Materials Interfaces*, *6*(21), 1900622. DOI: 10.1002/admi.201900622 PMID: 32313805

Bontempi, E., Sorrentino, G. P., Zanoletti, A., Alessandri, I., Depero, L. E., & Caneschi, A. (2021). Sustainable Materials and their Contribution to the Sustainable Development Goals (SDGs): A Critical Review Based on an Italian Example. *Molecules (Basel, Switzerland), 26*(5), 1407. DOI: 10.3390/molecules26051407 PMID: 33807763

Borella, L., Rozo, A., Perfetti, C., & Iorio, C. S. (2023). Characterization of Composite Freeze-Dried Aerogels with Simulant Lunar Regolith for Space Applications. *Materials (Basel), 16*(17), 5797. DOI: 10.3390/ma16175797 PMID: 37687489

Bouchaut, B., Asveld, L., Hanefeld, U., & Vlierboom, A. (2021). Value Conflicts in Designing for Safety: Distinguishing Applications of Safe-by-Design and the Inherent Safety Principles. *International Journal of Environmental Research and Public Health, 18*(4), 1963. DOI: 10.3390/ijerph18041963 PMID: 33670483

Caddeo, F., Loche, D., Casula, M. F., & Corrias, A. (2020). Growing CeO2 Nanoparticles Within the Nano-Porous Architecture of the SiO2 Aerogel. *Frontiers in Chemistry, 8*, 57. DOI: 10.3389/fchem.2020.00057 PMID: 32117882

Chen, Z., Fu, X., Liu, R., Song, Y., & Yin, X. (2023). Fabrication, Performance, and Potential Applications of MXene Composite Aerogels. *Nanomaterials (Basel, Switzerland), 13*(14), 2048. DOI: 10.3390/nano13142048 PMID: 37513059

Cristea, M., Ionita, D., & Iftime, M. M. (2020). Dynamic Mechanical Analysis Investigations of PLA-Based Renewable Materials: How Are They Useful? *Materials (Basel), 13*(22), 5302. DOI: 10.3390/ma13225302 PMID: 33238537

Filipovic, L., & Grasser, T. (2022). Special Issue on Miniaturized Transistors, Volume II. *Micromachines, 13*(4), 603. DOI: 10.3390/mi13040603 PMID: 35457908

Ganobjak, M., Brunner, S., Hofmann, J., Klar, V., Ledermann, M., Herzog, V., Kämpfen, B., Kilian, R., Wehdorn, M., & Wernery, J. (2023). Current Trends in Aerogel Use in Heritage Buildings: Case Studies from the Aerogel Architecture Award 2021. *Gels (Basel, Switzerland)*, *9*(10), 814. DOI: 10.3390/gels9100814 PMID: 37888387

Ghica, M. E., Mandinga, J. G. S., Linhares, T., Almeida, C. M. R., & Durães, L. (2023). Improvement of the Mechanical Properties of Silica Aerogels for Thermal Insulation Applications through a Combination of Aramid Nanofibres and Microfibres. *Gels (Basel, Switzerland)*, *9*(7), 535. DOI: 10.3390/gels9070535 PMID: 37504414

Grisolia, A., Dell'Olio, G., Spadafora, A., De Santo, M., Morelli, C., Leggio, A., & Pasqua, L. (2023). Hybrid Polymer-Silica Nanostructured Materials for Environmental Remediation. *Molecules (Basel, Switzerland)*, *28*(13), 5105. DOI: 10.3390/molecules28135105 PMID: 37446768

Guerra, F. D., Attia, M. F., Whitehead, D. C., & Alexis, F. (2018). Nanotechnology for Environmental Remediation: Materials and Applications. *Molecules (Basel, Switzerland)*, *23*(7), 1760. DOI: 10.3390/molecules23071760 PMID: 30021974

Guzzinati, G., Altantzis, T., Batuk, M., De Backer, A., Lumbeeck, G., Samaee, V., Batuk, D., Idrissi, H., Hadermann, J., Van Aert, S., Schryvers, D., Verbeeck, J., & Bals, S. (2018). Recent Advances in Transmission Electron Microscopy for Materials Science at the EMAT Lab of the University of Antwerp. *Materials (Basel)*, *11*(8), 1304. DOI: 10.3390/ma11081304 PMID: 30060556

Han, F., Li, M., Ye, H., & Zhang, G. (2021). Materials, Electrical Performance, Mechanisms, Applications, and Manufacturing Approaches for Flexible Strain Sensors. *Nanomaterials (Basel, Switzerland)*, *11*(5), 1220. DOI: 10.3390/nano11051220 PMID: 34063165

Horvat, G., Pantić, M., Knez, Ž., & Novak, Z. (2022). A Brief Evaluation of Pore Structure Determination for Bioaerogels. *Gels (Basel, Switzerland)*, *8*(7), 438. DOI: 10.3390/gels8070438 PMID: 35877523

Inobeme, A., Adetunji, C. O., Mathew, J. T., Ajai, I. A., & Jacob, J. O. (2024)b. Characterization of Plant-Based Nanobiofungicides. In book: Handbook of Agricultural Biotechnology, John Wiley & Sons, 81-97. DOI: DOI: 10.1002/9781119836544.ch5

Inobeme, A., Adetunji, C. O., Mathew, J. T., Okonkwo, S., Bamigboye, M. O., Ajai, A. I., Afoso, E., & Inobeme, J. (2023). Advanced nanotechnology for the degradation of persistent organic pollutants. In Shah, M. (Ed.), *Microbial Degradation and Detoxification of Pollutants* (pp. 51–72). De Gruyter., DOI: 10.1515/9783110743623-003

Inobeme, A., Mathew, J.T., Adetunji, C.O., Agbugui, M.O., Osarenren, E.J., Oti, C., Ngonso, B.F., Inobeme, J., Oyedolapo, M.B., Bernard, E. &Jibrin, N.A. (2024a). Biocomposites with graphene derivatives. In book: Advances in Biocomposites and their Applications, Elsevier Inc., 149-166. DOI: .DOI: 10.1016/B978-0-443-19074-2.00005-8

Inobeme, A., Shahnawaz, M., Adetunji, C. O., Mathew, J. T., & Adetuyi, B. O. Popoola, Oluwabukola Atinuke; Olaitan, Falana Yetunde; Akinbo, O., Kolawole, O. M., Oyewole, O. A., Eniola, K.I.T., Yerima, M. B. (2024)c. Primary Microplastic: Source, Uses, Transportation from Land to Marine Environment. In book: Microplastic Pollution. Springer, Singapore, 1, 23–33. DOI: DOI: 10.1007/978-981-99-8357-5_2

Inobeme, A., Shahnawaz, M., Adetunji, C. O., Mathew, J. T., Adetuyi, B. O., Popoola, O. A., Olaitan, F. Y., Akinbo, O., Kolawole, O. M., Oyewole, O. A., Eniola, K. I. T., & Yerima, B. M. (2024)d. Microplastic as a Pathogenic Vector to Cause Diseases in Marine Biota. In book: Microplastic Pollution. Springer, Singapore, 1, 321–349. DOI: DOI: 10.1007/978-981-99-8357-5_19

Jóźwiak, B., Dzido, G., Zor bski, E., Kolanowska, A., J drysiak, R., Dziadosz, J., Libera, M., Boncel, S., & Dzida, M. (2020). Remarkable Thermal Conductivity Enhancement in Carbon-Based Ionanofluids: Effect of Nanoparticle Morphology. *ACS Applied Materials & Interfaces*, *12*(34), 38113–38123. DOI: 10.1021/acsami.0c09752 PMID: 32649171

Juhász, L., Moldován, K., Gurikov, P., Liebner, F., Fábián, I., Kalmár, J., & Cserháti, C. (2021). False Morphology of Aerogels Caused by Gold Coating for SEM Imaging. *Polymers, 13*(4), 588. DOI: 10.3390/polym13040588 PMID: 33669181

Kamanina, O. A., Saverina, E. A., Rybochkin, P. V., Arlyapov, V. A., Vereshchagin, A. N., & Ananikov, V. P. (2022). Preparation of Hybrid Sol-Gel Materials Based on Living Cells of Microorganisms and Their Application in Nanotechnology. *Nanomaterials (Basel, Switzerland), 12*(7), 1086. DOI: 10.3390/nano12071086 PMID: 35407203

Khan, N. R., Sharmin, T., & Bin Rashid, A. (2023). Exploring the Versatility of Aerogels: Broad Applications in Biomedical Engineering, Astronautics, Energy Storage, Biosensing, and Current Progress. *Heliyon, 10*(1), e23102. DOI: 10.1016/j.heliyon.2023.e23102 PMID: 38163169

Lázár, I., & Menelaou, M. (2023). Editorial for the Special Issue: "Aerogel Hybrids and Nanocomposites". *Gels (Basel, Switzerland), 9*(10), 812. DOI: 10.3390/gels9100812 PMID: 37888385

Long, L. Y., Weng, Y. X., & Wang, Y. Z. (2018). Cellulose Aerogels: Synthesis, Applications, and Prospects. *Polymers, 10*(6), 623. DOI: 10.3390/polym10060623 PMID: 30966656

Maponya, T. C., Makgopa, K., Somo, T. R., & Modibane, K. D. (2022). Highlighting the Importance of Characterization Techniques Employed in Adsorption Using Metal-Organic Frameworks for Water Treatment. *Polymers, 14*(17), 3613. DOI: 10.3390/polym14173613 PMID: 36080689

Mathew, J. T., Inobeme, A., Adetuyi, B. O., Adetunji, C. O., Popoola, O. A., Olaitan, F. Y., Akinbo, O., Shahnawaz, M., Oyewole, O. A., Eniola, K. I. T., & Yerima, B. M. (2024)d. General Introduction of Microplastic: Uses, Types, and Generation. In book: Microplastic Pollution. Springer, Singapore, 1, 3-21. DOI: DOI: 10.1007/978-981-99-8357-5_1

Mathew, J. T., Inobeme, A., Azeh, Y., Musah, M., Abdulkadir, A., Shaba, E. Y., Etsuyankpa, M. B., Musa, T. S., Muhammad, A. I., Ismail, H., Kanwa, A. M., Mamman, A., & Hussaini, J. (2024). c. Carbon Capture and Storage Via Electrochemical and Bioelectrochemical Techniques: A Review. *Caliphate Journal of Science & Technology*, *2*(2), 141–158. DOI: 10.4314/cajost.v6i2.3

Mathew, J. T., Inobeme, A., Musah, M., Azeh, Y., Abdullahi, A., Shaba, E. Y., Salihu, A. M., Muhammad, E. B., Josiah, J. G., Jibrin, N. A., Ismail, H., Muhammad, A. I., Maurice, J., Mamman, A., & Ndamitso, M. M. (2024). a. A Critical Review of Green Approach on Wastewater Treatment Strategies. *Journal of Applied Science & Environmental Management*, *28*(2), 363–391. DOI: 10.4314/jasem.v28i2.9

Mathew, J. T., Musah, M., Azeh, Y., & Muhammed, M. (2023). a. Adsorptive Removal of Selected Toxic Metals from Pharmaceutical Wastewater using Fe_3O_4/ZnONanocomposite. *Dutse Journal of Pure and Applied Sciences*, *9*(4a), 236–248. DOI: 10.4314/dujopas.v9i4a.22

Mathew, J. T., Musah, M., Azeh, Y., & Muhammed, M. (2023). b. Kinetic Study of Heavy Metals Removal from Pharmaceutical Wastewater Using Geopolymer/Fe_3O_4Nanocomposite. *Bima Journal of Science and Technology*, *7*(4), 152–163. DOI: 10.56892/bima.v7i4.539

Mathew, J. T., Musah, M., Azeh, Y., & Muhammed, M. (2024). b. Development of Fe3O4 Nanoparticles for the Removal of Some Toxic Metals from Pharmaceutical Wastewater. [CaJoST]. *Caliphate Journal of Science & Technology*, *6*(1), 26–34. DOI: 10.4314/cajost.v6i1.4

Meti, P., Wang, Q., Mahadik, D. B., Lee, K. Y., Gong, Y. D., & Park, H. H. (2023). Evolutionary Progress of Silica Aerogels and Their Classification Based on Composition: An Overview. *Nanomaterials (Basel, Switzerland)*, *13*(9), 1498. DOI: 10.3390/nano13091498 PMID: 37177045

Niculescu, A. G., Tudorache, D. I., Bocioagă, M., Mihaiescu, D. E., Hadibarata, T., & Grumezescu, A. M. (2024). An Updated Overview of Silica Aerogel-Based Nanomaterials. *Nanomaterials (Basel, Switzerland)*, *14*(5), 469. DOI: 10.3390/nano14050469 PMID: 38470798

Noman, M. T., Amor, N., Ali, A., Petrik, S., Coufal, R., Adach, K., & Fijalkowski, M. (2021). Aerogels for Biomedical, Energy and Sensing Applications. *Gels (Basel, Switzerland)*, *7*(4), 264. DOI: 10.3390/gels7040264 PMID: 34940324

Ogbezode, J. E., Ezealigo, U. S., Bello, A., Anye, V. C., & Onwualu, A. P. (2023). A narrative review of the synthesis, characterization, and applications of iron oxide nanoparticles. *Discover Nano*, *18*(1), 125. DOI: 10.1186/s11671-023-03898-2 PMID: 37815643

Okonta, D. E. (2023). Investigating the impact of building materials on energy efficiency and indoor cooling in Nigerian homes. *Heliyon*, *9*(9), e20316. DOI: 10.1016/j.heliyon.2023.e20316 PMID: 37809702

Olorunsola, A., Adetunji, C. O., Olaniyan, O., Adetunji, J. B., Akinbo, K., Adetuyi, B. O., Inobeme, A., Mathew, J. T., & Ajenifujah-Solebo, S. O. (2024). Application of Nanotechnology for the Production of Biopesticides, Bioinsecticides, Bioherbicides, Mosquitoe Repellants and Biofungicides. In book: Handbook of Agricultural Biotechnology, John Wiley & Sons, 293-309. DOI: DOI: 10.1002/9781119836483.ch11

Owais, M., Shiverskii, A., Pal, A. K., Mahato, B., & Abaimov, S. G. (2022). Recent Studies on Thermally Conductive 3D Aerogels/Foams with the Segregated Nanofiller Framework. *Polymers*, *14*(22), 4796. DOI: 10.3390/polym14224796 PMID: 36432922

Parapouli, M., Vasileiadis, A., Afendra, A. S., & Hatziloukas, E. (2020). Saccharomyces cerevisiae and its industrial applications. *AIMS Microbiology*, *6*(1), 1–31. DOI: 10.3934/microbiol.2020001 PMID: 32226912

Paras, Y., Yadav, K., Kumar, P., Teja, D. R., Chakraborty, S., Chakraborty, M., Mohapatra, S. S., Sahoo, A., Chou, M. M. C., Liang, C.-T., & Hang, D.-R. (2022). A Review on Low-Dimensional Nanomaterials: Nanofabrication, Characterization and Applications. *Nanomaterials (Basel, Switzerland)*, *13*(1), 160. DOI: 10.3390/nano13010160 PMID: 36616070

Qi, P., Zhu, H., Borodich, F., & Peng, Q. (2023). A Review of the Mechanical Properties of Graphene Aerogel Materials: Experimental Measurements and Computer Simulations. *Materials (Basel)*, *16*(5), 1800. DOI: 10.3390/ma16051800 PMID: 36902915

Şahin, İ., Özbakır, Y., İnönü, Z., Ulker, Z., & Erkey, C. (2017). Kinetics of Supercritical Drying of Gels. *Gels (Basel, Switzerland)*, *4*(1), 3. DOI: 10.3390/gels4010003 PMID: 30674780

Sayam, A., Rahman, A. N. M. M., Rahman, M. S., Smriti, S. A., Ahmed, F., Rabbi, M. F., Hossain, M., & Faruque, M. O. (2022). A review on carbon fiber-reinforced hierarchical composites: Mechanical performance, manufacturing process, structural applications and allied challenges. *Carbon Letters*, *32*(5), 1173–1205. DOI: 10.1007/s42823-022-00358-2

Shah, K. W., & Huseien, G. F. (2020). Biomimetic Self-Healing Cementitious Construction Materials for Smart Buildings. *Biomimetics (Basel, Switzerland)*, *5*(4), 47. DOI: 10.3390/biomimetics5040047 PMID: 33049925

Shah, N., Rehan, T., Li, X., Tetik, H., Yang, G., Zhao, K., & Lin, D. (2021). Magnetic aerogel: An advanced material of high importance. *RSC Advances*, *11*(13), 7187–7204. DOI: 10.1039/D0RA10275J PMID: 35423256

Slepičková Kasálková, N., Slepička, P., & Švorčík, V.SlepičkováKasálková. (2021). Carbon Nanostructures, Nanolayers, and Their Composites. *Nanomaterials (Basel, Switzerland)*, *11*(9), 2368. DOI: 10.3390/nano11092368 PMID: 34578684

Speranza, G. (2021). Carbon Nanomaterials: Synthesis, Functionalization and Sensing Applications. *Nanomaterials (Basel, Switzerland)*, *11*(4), 967. DOI: 10.3390/nano11040967 PMID: 33918769

Suryanarayana, C. (2019). Mechanical Alloying: A Novel Technique to Synthesize Advanced Materials. *Research (Washington, D.C.)*, *4219812*, 2019/4219812. Advance online publication. DOI: 10.34133/2019/4219812 PMID: 31549063

Tarazona, N. A., Machatschek, R., Balcucho, J., Castro-Mayorga, J. L., Saldarriaga, J. F., & Lendlein, A. (2022). Opportunities and challenges for integrating the development of sustainable polymer materials within an international circular (bio)economy concept. *MRS Energy & Sustainability : a Review Journal*, *9*(1), 28–34. DOI: 10.1557/s43581-021-00015-7 PMID: 37521367

Uddin, M. A., Afroj, S., Hasan, T., Carr, C., Novoselov, K. S., & Karim, N. (2022). Environmental Impacts of Personal Protective Clothing Used to Combat COVID- 19. *Advanced Sustainable Systems*, *6*(1), 2100176. DOI: 10.1002/adsu.202100176 PMID: 34901387

Vrabič-Brodnjak, U. (2023). Hybrid Materials of Bio-Based Aerogels for Sustainable Packaging Solutions. *Gels (Basel, Switzerland)*, *10*(1), 27. DOI: 10.3390/gels10010027 PMID: 38247750

Wu, L., Li, Y., Fu, Z., & Su, B. L. (2020). Hierarchically structured porous materials: Synthesis strategies and applications in energy storage. *National Science Review*, *7*(11), 1667–1701. DOI: 10.1093/nsr/nwaa183 PMID: 34691502

Xuan, L., Ju, Z., Skonieczna, M., Zhou, P. K., & Huang, R. (2023). Nanoparticles-induced potential toxicity on human health: Applications, toxicity mechanisms, and evaluation models. *MedComm*, *4*(4), e327. DOI: 10.1002/mco2.327 PMID: 37457660

Chapter 3
Aerogels and Their Composites for Sensing Applications

Abel Inobeme
Edo State University, Uzairue, Nigeria

Stanley Okonkwo
https://orcid.org/0000-0001-5517-3193
Department of Chemistry, Osaka Kyoiku University, Osaka, Japan

Charles Adetunji
Edo State University, Uzairue, Nigeria

Jonathan Inobeme
Ahmadu Bello University, Zaria, Nigeria

Goddidit Esiro Enoyoze
Edo State University, Uzairue, Nigeria

Tawakalitu Kelani
Edo State University, Uzairue, Nigeria

ABSTRACT

Aerogels have garnered significant attention in recent years owing to their unique properties, such as ultra-low density, high surface area, and excellent thermal insulation. These characteristics make aerogels promising candidates for various applications, including sensing. This work presents a comprehensive review of aerogels and their composites utilized in sensing applications. We discuss the synthesis methods of aerogels and their functionalization techniques to tailor their sensing properties. Recent

DOI: 10.4018/979-8-3693-5971-6.ch003

advancements in the development of aerogel-based sensors for detecting various analytes are also discussed. Overall, aerogels and their composites hold great potential for advancing sensing technologies in diverse fields ranging from environmental monitoring to biomedical diagnostics.

5.1 INTRODUCTION

Aerogels, colloidal materials characterized by their ultra-lightweight and highly porous structures, have emerged as versatile platforms for a wide array of applications due to their remarkable physical and chemical properties. Among these applications, the integration of aerogels into sensing devices has garnered significant interest in recent years. This interest stems from the unique combination of properties that aerogels offer, including high surface area, low density, tunable porosity, and excellent thermal insulation (figure 1). These properties make aerogels ideal candidates for sensing applications, where high sensitivity, selectivity, and stability are crucial (Lázár et al. 2023).

Figure 1. Properties and starting materials for aerogel synthesis

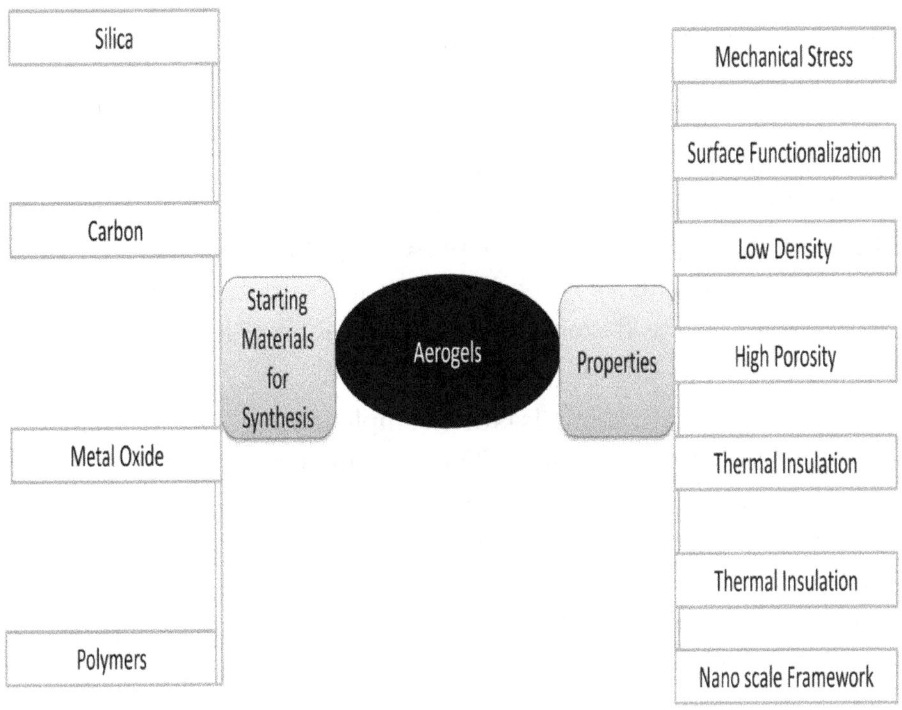

Aerogels are typically composed of interconnected nanoparticles or polymers, resulting in a highly porous structure with pores ranging from nanometers to micrometers in size. This porous network imparts aerogels with an extraordinarily high surface area, often exceeding several hundred square meters per gram. Consequently, aerogels exhibit exceptional adsorption and absorption capacities, making them suitable for sensing applications where interaction with analytes at the molecular level is essential (Paraskevopoulou et al. 2019).

Moreover, aerogels possess an extremely low bulk density, typically ranging from 0.001 to 0.5 g/cm^3, which contributes to their lightweight nature. This property is advantageous for portable sensing devices and applications where weight constraints are critical, such as wearable sensors or unmanned aerial vehicles (UAVs). Another notable feature of aerogels is their superior thermal insulation properties, stemming from their nanoscale pore structure and low density. This attribute enables aerogels to maintain temperature stability, making them suitable for sensing applications in harsh environments with fluctuating temperatures.

Aerogels can be synthesized through various methods, including sol-gel processes, supercritical drying, and freeze-drying. Sol-gel processes involve the hydrolysis and condensation of precursor materials, followed by gelation and subsequent removal of the solvent to obtain the aerogel structure. Supercritical drying involves replacing the liquid component of the gel with a supercritical fluid, preserving the gel's structure without collapse. Freeze-drying, or lyophilization, entails freezing the gel followed by sublimation of the frozen solvent under vacuum, leaving behind the aerogel (Wang et al. 2020).

Functionalization of aerogels is often employed to tailor their properties for specific sensing applications. This can involve surface modification with functional groups or incorporation of sensing elements such as nanoparticles, quantum dots, or polymers. These modifications enhance the aerogel's affinity towards target analytes, improve sensor response, and impart selectivity. Aerogels have been utilized in various sensing devices for the detection of gases, vapors, liquids, and biomolecules. Gas sensors based on aerogels have demonstrated high sensitivity and selectivity towards specific analytes such as volatile organic compounds (VOCs), ammonia, and carbon dioxide. Additionally, aerogel-based sensors have shown promise in environmental monitoring, industrial process control, and medical diagnostics (Aziz et al. 2022).

The incorporation of aerogels into composites further enhances the performance of sensing devices. Composite materials combining aerogels with polymers, nanomaterials, or functional additives exhibit synergistic effects, such as improved mechanical strength, enhanced sensitivity, and extended operational stability. These aerogel composites offer a versatile platform for designing customizable sensors with tailored properties to meet specific application requirements (Zhan et al. 2023).

In addition, aerogels and their composites represent promising platforms for sensing applications due to their unique combination of properties. Synthesis methods and functionalization techniques enable the customization of aerogel-based sensors for diverse applications, ranging from environmental monitoring to healthcare diagnostics. Continued research efforts aimed at advancing aerogel-based sensing technologies hold the potential to address critical challenges and drive innovation in sensing systems for various industrial, environmental, and biomedical applications (Almeida et al. 2024).

5.2 IMPORTANCE OF SENSING APPLICATIONS

Sensing applications play a pivotal role in modern society across a multitude of industries, ranging from healthcare and environmental monitoring to aerospace and automotive sectors. The ability to detect and quantify various physical, chemical, and biological parameters is crucial for ensuring safety, efficiency, and sustainability in numerous processes and systems. This article explores the significance of sensing applications and their impact on different aspects of our lives (Fascista, 2022).

Environmental Monitoring: One of the most critical areas where sensing applications are indispensable is environmental monitoring. Sensors are employed to measure parameters such as air and water quality, soil contamination, and greenhouse gas emissions. By providing real-time data on environmental conditions, sensing technologies help in identifying pollution sources, assessing environmental risks, and formulating effective mitigation strategies. This is crucial for safeguarding ecosystems, protecting public health, and promoting sustainable development (Ullo and Sinha, 2020; Mathew et al., 2023a).

Healthcare and Biomedical Diagnostics: In the healthcare sector, sensing applications play a vital role in disease diagnosis, patient monitoring, and drug development. Biosensors, for instance, enable the rapid and accurate detection of biomarkers, pathogens, and physiological parameters. These sensors are utilized in point-of-care diagnostics, wearable devices, and laboratory assays, facilitating early disease detection, personalized medicine, and remote patient monitoring. Additionally, sensing technologies contribute to advancements in drug delivery systems, biomedical imaging, and tissue engineering, thereby improving healthcare outcomes and enhancing quality of life (Yunus et al. 2023).

Industrial Process Control: Sensing applications are integral to industrial process control, where precise monitoring and regulation of various parameters are essential for optimizing production efficiency, ensuring product quality, and minimizing resource consumption. Sensors are employed in diverse industries such as manufacturing, chemical processing, and energy production to measure parameters such as temperature, pressure, flow rate, and chemical composition. By providing real-time feedback and control, sensing technologies enable automated process optimization, predictive maintenance, and compliance with regulatory standards, thereby enhancing productivity and reducing operational costs (Eifert et al. 2020; Inobeme et al., 2023).

Aerospace and Automotive Systems: In the aerospace and automotive sectors, sensing applications are critical for ensuring the safety, reliability, and performance of vehicles and aircraft. Sensors are utilized for monitoring engine parameters, detecting faults, and guiding navigation systems. In aerospace applications, sensors play a crucial role in flight control, avionics, and environmental monitoring, contributing to safe and efficient air travel. Similarly, in the automotive industry, sensors enable advanced driver assistance systems (ADAS), vehicle emissions monitoring, and predictive maintenance, enhancing road safety and environmental sustainability (Scott et al. 2022).

Smart Cities and Infrastructure: Sensing applications are instrumental in the development of smart cities and infrastructure, where the integration of sensor networks, Internet of Things (IoT) technologies, and data analytics enables efficient resource management, infrastructure maintenance, and urban planning. Sensors deployed in smart infrastructure systems monitor traffic flow, energy consumption, waste management, and public safety, facilitating real-time decision-making and improving the quality of urban life. By creating interconnected and data-driven urban environments, sensing technologies support sustainable development goals and enhance the resilience of cities to environmental and social challenges (Alahi et al. 2023).

In general, sensing applications play a vital role in addressing diverse challenges across various sectors, ranging from healthcare and environmental protection to industrial automation and urban development. By providing real-time data, enabling automation, and facilitating informed decision-making, sensing technologies contribute to improved efficiency, safety, and sustainability in our increasingly interconnected world. Continued advancements in sensor technology, data analytics, and communication systems are expected to further expand the scope and impact of sensing applications, driving innovation and fostering positive societal outcomes (Abdulmalek et al. 2022).

5.3 SYNTHESIS OF AEROGELS

Aerogels, ultralight materials with high porosity and low density, are synthesized through various methods, each offering unique advantages and limitations. The sol-gel process involves the hydrolysis and condensation of precursor materials, forming a gel that is then dried to remove the solvent, resulting in an aerogel structure. This method allows for precise control over composition and structure but can be time-consuming and prone to shrinkage. Supercritical drying replaces the liquid component of the gel with a supercritical fluid, preserving the gel's structure without collapse. It offers the advantage of maintaining the porous structure without shrinkage but requires specialized equipment and consumes high energy (Sheng et al. 2023).

Freeze-drying involves freezing the gel followed by sublimation of the frozen solvent under vacuum, leaving behind the aerogel. This method preserves the porous structure and is relatively simple, suitable for large-scale production, but it can be time-consuming and may result in pore collapse. Ambient pressure drying, on the other hand, involves drying the gel under ambient pressure conditions, often aided by solvent exchange or surface modification to prevent collapse. While simple and suitable for lab-scale synthesis, it offers limited control over pore structure and size and is prone to pore collapse without proper stabilization (Table 1) (Nowak and Jakubczyk,2020).

Table 1. Synthesis of aerogels

Synthesis Method	Description	Advantages	Disadvantages	References
Sol-Gel Process	Involves the hydrolysis and condensation of precursor materials, followed by gelation and subsequent removal of the solvent to obtain the aerogel structure.	Precise control over composition and structure, compatibility with various precursors.	Time-consuming process, requires careful control of reaction parameters, potential for shrinkage.	Kamanina et al. 2022
Supercritical Drying	Involves replacing the liquid component of the gel with a supercritical fluid, preserving the gel's structure without collapse.	Preserves the porous structure without shrinkage, allows for the formation of monolithic aerogels.	Requires specialized equipment and expertise, high energy consumption.	Guastaferro et al. 2021; Mathew et al., 2024a
Freeze-Drying	Involves freezing the gel followed by sublimation of the frozen solvent under vacuum, leaving behind the aerogel.	Preserves the porous structure, relatively simple process, suitable for large-scale production.	Long processing time, potential for pore collapse during drying, limited control over pore structure.	Katrilaka et al. 2023.
Ambient Pressure Drying	Involves drying the gel under ambient pressure conditions, typically aided by solvent exchange or surface modification to prevent collapse.	Simplicity and ease of operation, suitable for lab-scale synthesis.	Limited control over pore structure and size, prone to pore collapse without proper stabilization.	AbdulKhalil et al. 2020; Mathew et al., 2023b

5.4 FUNCTIONALIZATION OF AEROGELS

Functionalization of aerogels involves modifying their surface properties and incorporating additional functionalities to tailor their behavior for specific applications. This process enhances the performance of aerogels in areas such as catalysis, sensing, drug delivery, and environmental remediation. Here, we delve into the various methods and applications of aerogel functionalization (Bakhori *et al.* 2023). The functionalization of aerogels has various components (figure 2).

Figure 2. Functionalization of aerogels

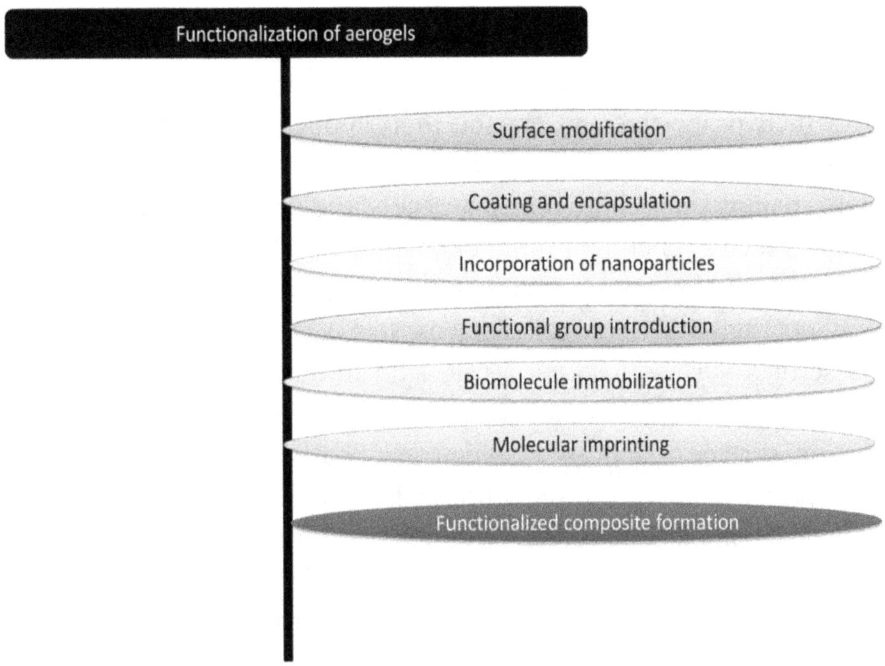

Surface Modification: Surface modification of aerogels is a widely employed strategy to enhance their properties and tailor them for specific applications. Chemical modification involves the covalent attachment of functional groups onto the aerogel surface. This can be achieved through reactions such as silanization, where organosilane compounds react with surface hydroxyl groups to form covalent bonds. Other chemical methods include grafting reactions and surface-initiated polymerization, which allow for precise control over surface chemistry and functionality. Physical modification, on the other hand, involves the adsorption or deposition of molecules onto the aerogel surface without forming covalent bonds. This can be achieved through techniques such as physical adsorption, electrostatic interaction, or layer-by-layer assembly. Physical modification methods are often simpler and more versatile than chemical methods, offering opportunities for introducing diverse functionalities onto the aerogel surface. Both chemical and physical modification techniques enable the customization of aerogels to meet the requirements of specific applications, such as sensing, catalysis, drug delivery, and environmental remediation (Franco et al. 2021).

Coating and Encapsulation: Coating aerogels with thin films or encapsulating them within protective layers is a valuable approach to improve their properties and expand their applicability. These coatings serve to enhance the stability, mechanical strength, and functionality of aerogels, making them more robust and versatile for various applications. Techniques such as chemical vapor deposition (CVD), atomic layer deposition (ALD), and layer-by-layer assembly (LbL) are commonly employed to apply coatings onto aerogel surfaces. In CVD, precursor gases react on the aerogel surface to form a thin film through chemical reactions. ALD, on the other hand, enables precise control over film thickness by alternating surface reactions between gas-phase precursors. Layer-by-layer assembly involves the sequential adsorption of oppositely charged polymers or nanoparticles onto the aerogel surface, allowing for precise control over coating composition and thickness. These coating techniques offer advantages such as uniform coverage, control over film thickness, and the ability to tailor coating properties. Coatings can provide protection against environmental factors, improve mechanical stability, enhance surface functionality, and even introduce new functionalities to aerogels, expanding their potential applications in fields such as catalysis, sensing, energy storage, and biomedical engineering (Frydenberg et al. 2022).

Incorporation of Nanoparticles: Incorporating nanoparticles into aerogel matrices is a versatile strategy to impart additional functionalities and tailor the properties of aerogels for specific applications. Nanoparticles, which can include metals, metal oxides, carbon-based materials, or polymers, offer unique properties such as catalytic activity, magnetic behavior, or enhanced adsorption capacity. Nanoparticles can be dispersed within the aerogel structure during synthesis or post-synthesis through techniques such as sol-gel chemistry or impregnation. This dispersion allows for intimate contact between nanoparticles and the aerogel matrix, leading to synergistic effects and improved performance (Chen et al. 2023).

Alternatively, nanoparticles can be immobilized onto the surface of aerogels through chemical bonding or physical adsorption. Chemical bonding involves covalent attachment of nanoparticles to functional groups on the aerogel surface, providing strong and stable interactions.

Physical adsorption relies on non-covalent forces such as electrostatic interactions or van der Waals forces to bind nanoparticles to the aerogel surface, offering simplicity and versatility. By incorporating nanoparticles into aerogel matrices, it is possible to enhance catalytic efficiency, introduce magnetic responsiveness, or improve adsorption capabilities, making functionalized aerogels valuable in applications such as catalysis, sensing, pollutant removal, and biomedical engineering (Thakur et al. 2021; Mathew et al., 2024b).

Functional Group Introduction: Introducing functional groups onto the aerogel surface is a fundamental method to tailor the chemical properties of aerogels and enable specific interactions with target molecules. Common functional groups include amino (-NH2), hydroxyl (-OH), carboxyl (-COOH), and thiol (-SH) groups, each imparting unique properties and functionalities. These functional groups can be introduced onto the aerogel surface through surface modification reactions or precursor functionalization during the synthesis process. Surface modification involves chemically grafting or attaching functional groups onto existing surface sites through reactions such as silanization or diazonium coupling. This approach allows for precise control over the density and distribution of functional groups on the aerogel surface (Meti et al. 2023).

Alternatively, functional groups can be introduced by incorporating precursor molecules containing the desired functional groups into the aerogel precursor solution. During the gelation and drying process, these precursor molecules react to form covalent bonds with the aerogel matrix, resulting in the incorporation of functional groups into the final aerogel structure. By introducing specific functional groups onto the aerogel surface, it becomes possible to enhance chemical reactivity, promote selective adsorption, or enable tailored interactions with target molecules. Functionalized aerogels find applications in various fields, including catalysis, sensing, drug delivery, and environmental remediation, where precise control over surface chemistry is essential for achieving desired performance (Keshavarz et al. 2021; Inobeme et al., 2024).

Templated Synthesis: Templated synthesis is a powerful approach to precisely control the pore structure of aerogels by using sacrificial templates. Various templates, including micelles, polymers, or colloidal particles, are employed to dictate the pore size, shape, and distribution within the aerogel matrix. During synthesis, the template is dispersed within the aerogel precursor solution, and upon gelation and removal of the template, well-defined pores are left behind in the aerogel structure. This technique allows for the creation of tailored properties and functionalities in aerogels, such as controlled porosity, specific surface area, and enhanced mechanical strength. Templated synthesis is particularly valuable in applications where pore structure plays a critical role, such as in catalysis, filtration, and energy storage, enabling the development of aerogels with optimized performance for specific applications (Song et al. 2022).

Biomolecule Immobilization: Immobilizing biomolecules onto aerogel surfaces unlocks a myriad of applications in biosensing, drug delivery, tissue engineering, and biocatalysis. Enzymes, antibodies, and DNA are examples of biomolecules that can be immobilized onto aerogels. This immobilization can be achieved through various methods including physical adsorption, covalent attachment, or affinity binding. Physical adsorption involves the non-covalent binding of biomolecules onto the aerogel surface, providing simplicity and versatility. Covalent attachment forms strong and stable bonds between biomolecules and the aerogel surface, ensuring robust immobilization. Affinity binding relies on specific interactions between biomolecules and ligands immobilized on the aerogel surface, enabling selective and reversible immobilization. By preserving the activity and specificity of biomolecules, immobilization onto aerogels enhances their performance in various applications, such as biosensors for detecting target analytes, drug delivery systems for targeted therapy, scaffolds for tissue regeneration, and biocatalysts for catalyzing biochemical reactions (Smith et al. 2020).

Molecular Imprinting: Molecular imprinting is a technique used to create selective recognition sites within the aerogel matrix by templating with target molecules or their analogs. During synthesis, the target molecules are incorporated into the aerogel precursor solution, which forms a molecular imprint around them as the aerogel solidifies. Subsequent removal of the template molecules leaves behind cavities or binding sites within the aerogel that are complementary in shape and chemical functionality to the template molecules. This results in molecularly imprinted aerogels with specific affinity towards the target molecules, enabling selective binding and detection in sensing and separation applications. Molecularly imprinted aerogels have shown promise in various fields, including environmental monitoring, pharmaceutical analysis, and chemical sensing, where the ability to selectively recognize and capture specific analytes is crucial for accurate detection and separation processes (Fresco-Cala et al. 2020).

Functional Composite Formation: Forming composites by integrating aerogels into matrices of polymers, ceramics, or other materials offers a synergistic enhancement of properties and functionalities. These functional composites leverage the unique characteristics of aerogels, such as high porosity and low density, with the mechanical strength, flexibility, or conductivity of the host matrix. The combination of these materials extends the range of applications for aerogels, enabling improved thermal insulation, enhanced mechanical stability, or increased electrical conductivity, depending on the specific requirements. By tailoring the composition and structure of the composite, it is possible to achieve desired properties for applications in areas such as aerospace, construction, energy storage, and electronics. The versatility and tunability of aerogel-based composites make them valuable materials for addressing diverse challenges across various industries (Jin et al. 2023).

5.4.1 Applications of Functionalized Aerogels

Functionalized aerogels find applications in diverse fields, including catalysis, sensing, drug delivery, environmental remediation, energy storage, and insulation. For example, catalytically active aerogels functionalized with metal nanoparticles exhibit high efficiency in catalytic reactions. Similarly, aerogel-based sensors functionalized with selective receptors demonstrate excellent sensitivity and specificity for detecting target analytes. Functionalized aerogels also serve as drug delivery vehicles, providing controlled release and targeted delivery of therapeutics. In environmental applications, functionalized aerogels are utilized for adsorption of pollutants, filtration of contaminants, and remediation of contaminated sites. Additionally, functionalized aerogels are explored for energy storage in batteries and supercapacitors, as well as for thermal insulation in buildings and aerospace applications (Budtova et al. 2020).

Moreover, the functionalization of aerogels enables the customization of their properties and functionalities for specific applications, expanding their utility across various industries. By tailoring surface chemistry, incorporating nanoparticles, and forming functional composites, functionalized aerogels offer enhanced performance and versatility in catalysis, sensing, drug delivery, environmental remediation, and beyond. Continued research in aerogel functionalization holds the promise of further advancing these materials and unlocking new opportunities for innovation and application development (Sozcu et al. 2023).

5.5 AEROGEL-BASED SENSORS

Aerogel-based sensors encompass various types with distinctive sensing mechanisms and applications (**Table 2**). Gas sensors rely on conductivity changes induced by gas interaction, suitable for environmental monitoring and industrial safety (figure 3). Biosensors employ biomolecular recognition for detecting biomolecules, essential in medical diagnostics and food safety. Chemical sensors detect chemical species through reactions or adsorption, valuable for environmental monitoring and process control. Optical sensors utilize light absorption or emission for high-sensitivity, non-contact measurement, applied in biomedical imaging and environmental sensing. Pressure sensors utilize mechanical deformation for detecting pressure variations, beneficial for industrial automation and wearable devices. The unique properties of aerogels, such as high surface area and porosity, make them versatile platforms for sensor development across diverse fields (Gatou et al. 2023).

Figure 3. Sensing mechanisms in aerogels

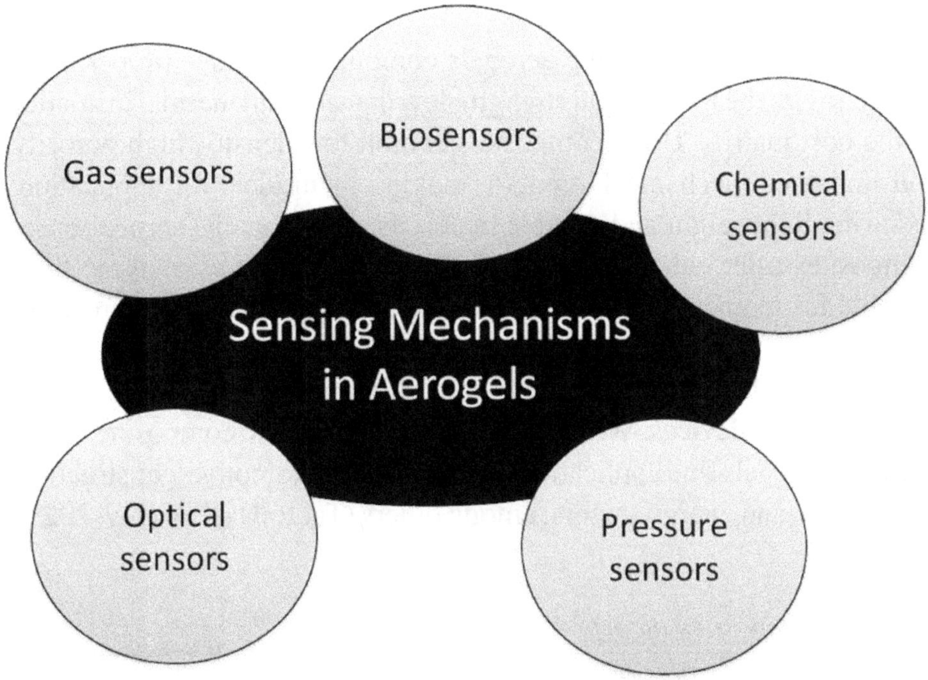

Table 2. Aerogel-based sensors

Sensor Type	Sensing Mechanism	Target Analyte	Advantages	Applications	References
Gas Sensors	Conductivity changes	Various gases	High sensitivity, fast response time	Environmental monitoring, industrial safety	Yan et al. 2019 Adetunji et al., 2023
Biosensors	Biomolecular recognition	Biomolecules	Specificity, biocompatibility	Medical diagnostics, food safety	Naresh and Lee, 2021.
Chemical Sensors	Chemical reaction or adsorption	Chemical species	Selectivity, wide detection range	Environmental monitoring, process control	Baranwal et al. 2022.
Optical Sensors	Light absorption or emission	Various analytes	High sensitivity, non-contact measurement	Biomedical imaging, environmental sensing	Butt et al. 2022.
Pressure Sensors	Mechanical deformation	Pressure	High sensitivity, miniaturization	Industrial automation, wearable devices	Zhang and Zhang, 2023.

5.6 COMPOSITES INCORPORATING AEROGELS

Composites incorporating aerogels combine the unique properties of aerogels with the mechanical strength, low density, and thermal insulation of the host matrix. These composites exhibit low density, high porosity, and improved mechanical strength, making them ideal for applications requiring lightweight and durable materials. Additionally, aerogel-based composites offer enhanced thermal insulation properties, making them suitable for temperature control and insulation applications. Depending on the type of aerogel and host matrix, these composites can also exhibit electrical conductivity, expanding their utility in electronics, sensors, and energy storage devices. With their versatile properties, composites incorporating aerogels find applications in aerospace, automotive, construction, electronics, and energy sectors, among others (Table 3) (Kim *et al.* 2021).

Table 3. Composites incorporating aerogels

Property	Description	Advantages	Applications	References
Low Density	Composites incorporating aerogels exhibit low density due to the lightweight nature of aerogels, making them suitable for weight-sensitive applications.	Reduced weight, improved fuel efficiency.	Aerospace, automotive, lightweight structures.	Jin et al. 2023.
High Porosity	The inherent porosity of aerogels enhances the composite's thermal insulation properties, making them suitable for applications requiring temperature control or insulation.	Enhanced thermal insulation, energy efficiency.	Building materials, thermal packaging, cryogenic applications.	Fijalkowski et al. 2023.
Mechanical Strength	Incorporating aerogels into the host matrix can improve the mechanical strength and toughness of the composite, reducing the risk of fracture or failure under mechanical loads.	Enhanced mechanical properties, increased durability.	Structural components, load-bearing applications.	Ghica et al. 2023.
Electrical Conductivity	Depending on the type of aerogel and host matrix, composites incorporating aerogels can exhibit electrical conductivity, making them suitable for applications in electronics, sensors, and energy storage devices.	Enhanced electrical properties, conductivity.	Electronics, sensors, energy storage devices.	Chen et al. 2023.

5.7 ADVANTAGES OF AEROGEL-BASED SENSORS

Aerogel-based sensors offer a range of advantages that make them attractive for various sensing applications across diverse industries. Here are some main advantages of aerogel-based sensors:

High Sensitivity: Aerogels possess an exceptionally high surface area and porosity, providing ample active sites for interactions with target analytes. This high surface-to-volume ratio enhances the sensitivity of aerogel-based sensors, enabling the detection of trace amounts of analytes with high precision and accuracy (Noman et al. 2021).

Selectivity: Functionalization of aerogel surfaces allows for specific interactions with target molecules, leading to enhanced selectivity in sensing. By incorporating specific receptors or functional groups onto the aerogel surface, sensors can discriminate between different analytes, even in complex sample matrices (Basso et al. 2023).

Rapid Response Time: The porous structure of aerogels facilitates rapid diffusion of analytes into the sensing material, resulting in fast response times. This enables real-time monitoring and detection of changes in analyte concentrations, crucial for applications requiring timely responses, such as environmental monitoring or process control (Yu et al. 2023).

Versatility: Aerogel-based sensors can be tailored to detect a wide range of analytes, including gases, vapors, liquids, and biomolecules. This versatility allows for the adaptation of aerogel sensors to various industries and applications, from environmental monitoring to biomedical diagnostics (Nita et al. 2020).

Stability and Durability: Aerogels exhibit excellent stability and durability, even under harsh environmental conditions. This stability ensures the long-term performance and reliability of aerogel-based sensors, making them suitable for continuous monitoring applications in challenging environments (Noman et al. 2021).

Miniaturization: The lightweight and compact nature of aerogels allows for the miniaturization of sensors, enabling the development of portable and wearable devices. Miniaturized aerogel-based sensors offer opportunities for on-site and in-situ monitoring, facilitating rapid deployment and data collection in remote or inaccessible locations (Almeida et al. 2024).

Biocompatibility: Aerogels are inherently biocompatible materials, making them suitable for biomedical and healthcare applications. Aerogel-based sensors can be integrated into implantable devices, wearable biosensors, or diagnostic tools for non-invasive monitoring of physiological parameters or biomarkers (Bakhori et al. 2023).

Low Power Consumption: Aerogel-based sensors typically require low power consumption, making them energy-efficient and cost-effective for long-term deployment. This feature is advantageous for remote sensing applications or battery-powered devices where power constraints are a concern (He et al. 2023).

Resistance to Fouling: The porous structure of aerogels minimizes fouling effects by allowing analytes to penetrate into the material, reducing surface adsorption and buildup of contaminants. This resistance to fouling ensures the reliability and accuracy of aerogel-based sensors over extended periods of operation (Sharma et al. 2023).

Scalability: Aerogel synthesis processes can be scaled up to produce large quantities of sensor materials, allowing for cost-effective manufacturing and widespread deployment. This scalability is essential for commercialization and mass production of aerogel-based sensors for industrial, environmental, and consumer applications (Nguyen et al. 2022).

In general, aerogel-based sensors offer a unique combination of high sensitivity, selectivity, rapid response time, versatility, stability, and miniaturization, making them attractive for a wide range of sensing applications. With ongoing advancements in aerogel synthesis, functionalization techniques, and sensor integration technologies, aerogel-based sensors hold great promise for addressing emerging challenges and driving innovation in fields such as environmental monitoring, healthcare, safety, and beyond (Shirsat & Hianik, 2023).

5.8 CHALLENGES IN PRACTICAL APPLICATIONS

While aerogel-based sensors offer numerous advantages, several challenges must be addressed for their practical implementation across various applications. Here are some key challenges:

Sensitivity and Selectivity Optimization: Achieving optimal sensitivity and selectivity is crucial for accurate and reliable detection of target analytes. However, enhancing sensitivity without compromising selectivity and vice versa remains a challenge. Balancing these parameters often requires careful optimization of sensor materials, surface functionalization strategies, and detection techniques (Sathish & Shen,2021).

Stability and Longevity: Maintaining sensor performance over extended periods, especially in harsh environments, is essential for practical applications. Aerogel-based sensors may experience degradation or loss of functionality due to environmental factors such as temperature fluctuations, humidity, chemical exposure, or mechanical stress. Developing robust sensor architectures and protective coatings to enhance stability and longevity is critical (Seesaard & Wongchoosuk, 2023).

Reproducibility and Consistency: Ensuring reproducibility and consistency in sensor performance is essential for widespread adoption and commercialization. Variability in aerogel properties, fabrication processes, or surface functionalization techniques can lead to inconsistent sensor responses. Standardizing manufacturing processes, quality control measures, and calibration protocols can address these challenges (Pradhan et al. 2020).

Interference and Cross-Sensitivity: Aerogel-based sensors may exhibit interference or cross-sensitivity to other analytes or environmental factors present in complex sample matrices. Discriminating between target analytes and interfering species requires robust sensor designs, advanced signal processing algorithms, and selective recognition elements. Minimizing cross-reactivity and enhancing specificity are ongoing challenges in sensor development (Kang et al. 2021).

Miniaturization and Integration: While miniaturization enables the development of portable and wearable sensor devices, integrating aerogel-based sensors into compact platforms presents engineering challenges. Designing miniaturized sensor arrays, signal transduction mechanisms, and power sources while maintaining sensor performance and reliability requires interdisciplinary expertise in materials science, electronics, and microfabrication (Jia et al. 2022).

Cost and Scalability: Cost-effective manufacturing and scalability are critical for widespread adoption of aerogel-based sensors in commercial applications. High production costs associated with aerogel synthesis, functionalization, and sensor fabrication may limit their accessibility. Developing scalable production methods, leveraging low-cost materials, and optimizing sensor designs for mass production can address these challenges (Sozcu et al. 2023).

Biocompatibility and Regulatory Approval: For biomedical and healthcare applications, ensuring the biocompatibility and regulatory compliance of aerogel-based sensors is paramount. Biocompatibility assessments, biostability studies, and adherence to regulatory standards are essential steps in the development and commercialization process. Addressing safety concerns and obtaining regulatory approval are significant challenges in translating aerogel-based sensors from research to clinical use (Zheng et al. 2020).

Real-World Testing and Validation: Validating sensor performance under real-world conditions is critical to assessing their reliability and efficacy in practical applications. Conducting field trials, environmental testing, and performance evaluations in diverse settings can provide valuable insights into sensor behavior and identify potential challenges or limitations. Collaborating with end-users, industry partners, and regulatory agencies is essential for conducting comprehensive validation studies (Mazzà et al. 2021).

Data Interpretation and Analysis: Analyzing and interpreting sensor data accurately and efficiently is essential for extracting meaningful insights and making informed decisions. Developing advanced data analytics algorithms, machine learning models, and sensor fusion techniques to process, interpret, and visualize complex datasets from aerogel-based sensors is an ongoing research area (Batko & Ślęzak, 2022).

User Acceptance and Adoption: User acceptance and adoption of aerogel-based sensors depend on factors such as usability, reliability, affordability, and perceived value. Educating end-users, addressing usability concerns, and demonstrating the benefits of aerogel-based sensors in real-world applications are essential for fostering trust and driving widespread adoption (Li et al. 2023).

In addition, addressing these challenges requires interdisciplinary collaboration, innovative approaches, and continuous advancements in materials science, sensor technology, and system integration. Overcoming these hurdles will unlock the full potential of aerogel-based sensors and enable their widespread deployment in diverse fields, ranging from environmental monitoring and healthcare to industrial process control and consumer electronics (Bakhori et al. 2023).

5.9 CONCLUSION

In general, aerogels and their composites hold tremendous promise for sensing applications across various industries. Their unique combination of properties, including high surface area, porosity, low density, and customizable surface chemistry, makes them ideal platforms for developing advanced sensors with enhanced sensitivity, selectivity, and versatility. Through innovative synthesis methods, surface functionalization techniques, and integration into composite materials, aerogels can address complex sensing challenges in fields such as environmental monitoring, healthcare, safety, and beyond. Composites incorporating aerogels offer synergistic advantages, combining the unique characteristics of aerogels with the mechanical strength, flexibility, or conductivity of the host matrix, thereby expanding the range of sensing applications.

However, practical implementation of aerogel-based sensors still faces challenges such as sensitivity optimization, stability enhancement, miniaturization, and cost-effectiveness. Addressing these challenges requires continued research and development efforts, interdisciplinary collaboration, and advancements in materials science, sensor technology, and system integration. With ongoing innovations and advancements, aerogel-based sensors are poised to revolutionize sensing technologies, enabling rapid and accurate detection of target analytes, enhancing environmental monitoring capabilities, improving healthcare diagnostics, and driving advancements in various industries.

REFERENCES

Abdul Khalil, H. P. S., Adnan, A. S., Yahya, E. B., Olaiya, N. G., Safrida, S., Hossain, M. S., Balakrishnan, V., Gopakumar, D. A., Abdullah, C. K., Oyekanmi, A. A., & Pasquini, D. (2020). A Review on Plant Cellulose Nanofibre-Based Aerogels for Biomedical Applications. *Polymers, 12*(8), 1759. DOI: 10.3390/polym12081759 PMID: 32781602

Abdulmalek, S., Nasir, A., Jabbar, W. A., Almuhaya, M. A. M., Bairagi, A. K., Khan, M. A., & Kee, S. H. (2022). IoT-Based Healthcare-Monitoring System towards Improving Quality of Life: A Review. *Healthcare (Basel), 10*(10), 1993. DOI: 10.3390/healthcare10101993 PMID: 36292441

Adetunji, C. O., Mathew, J. T., Inobeme, A., Taiwo, O., Olaniyan, O., Shakira, G., & Wadazani, D.(2023).Application of Biosensors for Detection and Monitoring of Water Quality.In book: Organic and Inorganic Materials Based Sensors, John Wiley & Sons, Inc., 1065-1077, DOI: 10.3390/healthcare10101993 PMID: 36292441

Alahi, M. E. E., Sukkuea, A., Tina, F. W., Nag, A., Kurdthongmee, W., Suwannarat, K., & Mukhopadhyay, S. C. (2023). Integration of IoT-Enabled Technologies and Artificial Intelligence (AI) for Smart City Scenario: Recent Advancements and Future Trends. *Sensors (Basel), 23*(11), 5206. DOI: 10.3390/s23115206 PMID: 37299934

Almeida, C. M. R., Merillas, B., & Pontinha, A. D. R. (2024). Trends on Aerogel-Based Biosensors for Medical Applications: An Overview. *International Journal of Molecular Sciences, 25*(2), 1309. DOI: 10.3390/ijms25021309 PMID: 38279307

Aziz, T., Farid, A., Haq, F., Kiran, M., Ullah, A., Zhang, K., Li, C., Ghazanfar, S., Sun, H., Ullah, R., Ali, A., Muzammal, M., Shah, M., Akhtar, N., Selim, S., Hagagy, N., Samy, M., & Al Jaouni, S. K. (2022). A Review on the Modification of Cellulose and Its Applications. *Polymers, 14*(15), 3206. DOI: 10.3390/polym14153206 PMID: 35956720

Bakhori, N. M., Ismail, Z., Hassan, M. Z., & Dolah, R. (2023). Emerging Trends in Nanotechnology: Aerogel-Based Materials for Biomedical Applications. *Nanomaterials (Basel, Switzerland), 13*(6), 1063. DOI: 10.3390/nano13061063 PMID: 36985957

Baranwal, J., Barse, B., Gatto, G., Broncova, G., & Kumar, A. (2022). Electrochemical Sensors and Their Applications: A Review. *Chemosensors (Basel, Switzerland), 10*(9), 363. DOI: 10.3390/chemosensors10090363

Basso, C. R., Crulhas, B. P., Castro, G. R., & Pedrosa, V. A. (2023). Recent Advances in Functional Nanomaterials for Diagnostic and Sensing Using Self-Assembled Monolayers. *International Journal of Molecular Sciences, 24*(13), 10819. DOI: 10.3390/ijms241310819 PMID: 37445998

Batko, K., & Ślęzak, A. (2022). The use of Big Data Analytics in healthcare. *Journal of Big Data, 9*(1), 3. DOI: 10.1186/s40537-021-00553-4 PMID: 35013701

Budtova, T., Aguilera, D. A., Beluns, S., Berglund, L., Chartier, C., Espinosa, E., Gaidukovs, S., Klimek-Kopyra, A., Kmita, A., Lachowicz, D., Liebner, F., Platnieks, O., Rodríguez, A., Navarro, L. K. T., Zou, F., & Buwalda, S. J. (2020). Biorefinery Approach for Aerogels. *Polymers, 12*(12), 2779. DOI: 10.3390/polym12122779 PMID: 33255498

Butt, M. A., Voronkov, G. S., Grakhova, E. P., Kutluyarov, R. V., Kazanskiy, N. L., & Khonina, S. N. (2022). Environmental Monitoring: A Comprehensive Review on Optical Waveguide and Fiber-Based Sensors. *Biosensors (Basel), 12*(11), 1038. DOI: 10.3390/bios12111038 PMID: 36421155

Chen, Z., Fu, X., Liu, R., Song, Y., & Yin, X. (2023). Fabrication, Performance, and Potential Applications of MXene Composite Aerogels. *Nanomaterials (Basel, Switzerland), 13*(14), 2048. DOI: 10.3390/nano13142048 PMID: 37513059

Eifert, T., Eisen, K., Maiwald, M., & Herwig, C. (2020). Current and future requirements to industrial analytical infrastructure-part 2: Smart sensors. *Analytical and Bioanalytical Chemistry, 412*(9), 2037–2045. DOI: 10.1007/s00216-020-02421-1 PMID: 32055909

Fascista, A. (2022). Toward Integrated Large-Scale Environmental Monitoring Using WSN/UAV/Crowdsensing: A Review of Applications, Signal Processing, and Future Perspectives. *Sensors (Basel)*, *22*(5), 1824. DOI: 10.3390/s22051824 PMID: 35270970

Fijalkowski, M., Coufal, R., Ali, A., Adach, K., Petrik, S., Bu, H., & Karl, C. W. (2023). Flexible Hybrid and Single-Component Aerogels: Synthesis, Characterization, and Applications. *Langmuir*, *39*(47), 16760–16775. DOI: 10.1021/acs.langmuir.3c01811 PMID: 37955990

Franco, P., Cardea, S., Tabernero, A., & De Marco, I. (2021). Porous Aerogels and Adsorption of Pollutants from Water and Air: A Review. *Molecules (Basel, Switzerland)*, *26*(15), 4440. DOI: 10.3390/molecules26154440 PMID: 34361593

Fresco-Cala, B., Batista, A. D., & Cárdenas, S. (2020). Molecularly Imprinted Polymer Micro- and Nano-Particles. A review. *Molecules (Basel, Switzerland)*, *25*(20), 4740. DOI: 10.3390/molecules25204740 PMID: 33076552

Frydenberg, T., Weinell, C. E., Dam-Johansen, K., Wallström, E., & Kiil, S. (2022). Characterization and Release Mechanisms of Aerogel-Encapsulated Biocide Crystals for Low-Loading and High-Utilization Antifouling Coatings. *ACS Omega*, *7*(39), 34824–34838. DOI: 10.1021/acsomega.2c03133 PMID: 36211059

Gatou, M. A., Vagena, I. A., Lagopati, N., Pippa, N., Gazouli, M., & Pavlatou, E. A. (2023). Functional MOF-Based Materials for Environmental and Biomedical Applications: A Critical Review. *Nanomaterials (Basel, Switzerland)*, *13*(15), 2224. DOI: 10.3390/nano13152224 PMID: 37570542

Ghica, M. E., Mandinga, J. G. S., Linhares, T., Almeida, C. M. R., & Durães, L. (2023). Improvement of the Mechanical Properties of Silica Aerogels for Thermal Insulation Applications through a Combination of Aramid Nanofibres and Microfibres. *Gels (Basel, Switzerland)*, *9*(7), 535. DOI: 10.3390/gels9070535 PMID: 37504414

Guastaferro, M., Reverchon, E., & Baldino, L. (2021). Polysaccharide-Based Aerogel Production for Biomedical Applications: A Comparative Review. *Materials (Basel), 14*(7), 1631. DOI: 10.3390/ma14071631 PMID: 33810582

He, D., Cui, Y., Ming, F., & Wu, W. (2023). Advancements in Passive Wireless Sensors, Materials, Devices, and Applications. *Sensors (Basel), 23*(19), 8200. DOI: 10.3390/s23198200 PMID: 37837030

Inobeme, A., Adetunji, C. O., Mathew, J T., Ajai, A. I., Inobeme, J., Bamigboye, M. O., Onyeaku, S., Maliki, M., Eziukwu, C. &Tawa, K. (2023). Nanotechnology for Bioremediation of Heavy Metals. In book: Microbial Technologies in Industrial Wastewater Treatment. Springer Nature Singapore, 19-30. .DOI: 10.1007/978-981-99-2435-6_2

Inobeme, A., Mathew, J.T., Adetunji, C.O., Agbugui, M.O., Osarenren, E.J., Oti, C., Ngonso, B.F., Inobeme, J., Oyedolapo, M.B., Bernard, E. &Jibrin, N.A. (2024). Biocomposites with graphene derivatives. In book: Advances in Biocomposites and their Applications, Elsevier Inc., 149-166. DOI: DOI: 10.1016/B978-0-443-19074-2.00005-8

Jia, S., Gao, H., Xue, Z., & Meng, X. (2022). Recent Advances in Multifunctional Wearable Sensors and Systems: Design, Fabrication, and Applications. *Biosensors (Basel), 12*(11), 1057. DOI: 10.3390/bios12111057 PMID: 36421175

Jin, R., Zhou, Z., Liu, J., Shi, B., Zhou, N., Wang, X., Jia, X., Guo, D., & Xu, B. (2023). Aerogels for Thermal Protection and Their Application in Aerospace. *Gels (Basel, Switzerland), 9*(8), 606. DOI: 10.3390/gels9080606 PMID: 37623061

Kamanina, O. A., Saverina, E. A., Rybochkin, P. V., Arlyapov, V. A., Vereshchagin, A. N., & Ananikov, V. P. (2022). Preparation of Hybrid Sol-Gel Materials Based on Living Cells of Microorganisms and Their Application in Nanotechnology. *Nanomaterials (Basel, Switzerland), 12*(7), 1086. DOI: 10.3390/nano12071086 PMID: 35407203

Kang, X., Yip, S., Meng, Y., Wang, W., Li, D., Liu, C., & Ho, J. C. (2021). High-performance electrically transduced hazardous gas sensors based on low-dimensional nanomaterials. *Nanoscale Advances, 3*(22), 6254–6270. DOI: 10.1039/D1NA00433F PMID: 36133491

Katrilaka, C., Karipidou, N., Petrou, N., Manglaris, C., Katrilakas, G., Tzavellas, A. N., Pitou, M., Tsiridis, E. E., Choli-Papadopoulou, T., & Aggeli, A. (2023). Freeze-Drying Process for the Fabrication of Collagen-Based Sponges as Medical Devices in Biomedical Engineering. *Materials (Basel), 16*(12), 4425. DOI: 10.3390/ma16124425 PMID: 37374608

Keshavarz, L., Ghaani, M. R., & English, N. J. (2021). The Importance of Precursors and Modification Groups of Aerogels in CO2 Capture. *Molecules (Basel, Switzerland), 26*(16), 5023. DOI: 10.3390/molecules26165023 PMID: 34443610

Kim, J.-H., Ahn, J.-H., Kim, J.-D., Lee, D.-H., Kim, S.-K., & Lee, J.-M. (2021). Influence of Silica-Aerogel on Mechanical Characteristics of Polyurethane-Based Composites: Thermal Conductivity and Strength. *Materials (Basel), 14*(7), 1790. DOI: 10.3390/ma14071790 PMID: 33916354

Lázár, I., Čelko, L., & Menelaou, M. (2023). Aerogel-Based Materials in Bone and Cartilage Tissue Engineering-A Review with Future Implications. *Gels (Basel, Switzerland), 9*(9), 746. DOI: 10.3390/gels9090746 PMID: 37754427

Li, S., Li, H., Lu, Y., Zhou, M., Jiang, S., Du, X., & Guo, C. (2023). Advanced Textile-Based Wearable Biosensors for Healthcare Monitoring. *Biosensors (Basel), 13*(10), 909. DOI: 10.3390/bios13100909 PMID: 37887102

Mathew, J. T., Inobeme, A., Abdulkadir, A., Etsuyankpa, M. B., Shaba, E. Y., Musah, M., Azeh, Y., Adetunji, C. O., Ibrahim, A. M., Tanko, E., & Mamman, A. (2024). a. Properties and Characteristics of Nanocoatings for Medicinal Applications. In book: Sustainable Approach to Protective Nanocoatings. IGI Global; publishing Tomorrow's. *Advances in Chemical and Materials Engineering*, •••, 203–227. DOI: 10.4018/979-8-3693-3136-1.ch007

Mathew, J. T., Inobeme, A., Musah, M., Azeh, Y., Abdullahi, A., Shaba, E. Y., Salihu, A. M., Muhammad, E. B., Josiah, J. G., Jibrin, N. A., Ismail, H., Muhammad, A. I., Maurice, J., Mamman, A., & Ndamitso, M. M. (2024). b. A Critical Review of Green Approach on Wastewater Treatment Strategies. *Journal of Applied Science & Environmental Management, 28*(2), 363–391. DOI: 10.4314/jasem.v28i2.9

Mathew, J. T., Musah, M., Azeh, Y., & Muhammed, M. (2023). a. Kinetic Study of Heavy Metals Removal from Pharmaceutical Wastewater Using Geopolymer/Fe_3O_4Nanocomposite. *Bima Journal of Science and Technology*, *7*(4), 152–163. DOI: 10.56892/bima.v7i4.539

Mathew, J. T., Musah, M., Azeh, Y., & Muhammed, M. (2023). b. Adsorptive Removal of Selected Toxic Metals from Pharmaceutical Wastewater using Fe_3O_4/ZnONanocomposite. *Dutse Journal of Pure and Applied Sciences*, *9*(4a), 236–248. DOI: 10.4314/dujopas.v9i4a.22

Mazzà, C., Alcock, L., Aminian, K., Becker, C., Bertuletti, S., Bonci, T., Brown, P., Brozgol, M., Buckley, E., Carsin, A. E., Caruso, M., Caulfield, B., Cereatti, A., Chiari, L., Chynkiamis, N., Ciravegna, F., Del Din, S., Eskofier, B., Evers, J., & Rochester, L. (2021). Technical validation of real-world monitoring of gait: A multicentric observational study. *BMJ Open*, *11*(12), e050785. DOI: 10.1136/bmjopen-2021-050785 PMID: 34857567

Meti, P., Wang, Q., Mahadik, D. B., Lee, K. Y., Gong, Y. D., & Park, H. H. (2023). Evolutionary Progress of Silica Aerogels and Their Classification Based on Composition: An Overview. *Nanomaterials (Basel, Switzerland)*, *13*(9), 1498. DOI: 10.3390/nano13091498 PMID: 37177045

Naresh, V., & Lee, N. (2021). A Review on Biosensors and Recent Development of Nanostructured Materials-Enabled Biosensors. *Sensors (Basel)*, *21*(4), 1109. DOI: 10.3390/s21041109 PMID: 33562639

Nguyen, P. T. T., Do, N. H. N., Goh, X. Y., Goh, C. J., Ong, R. H., Le, P. K., Phan-Thien, N., & Duong, H. M. (2022). Recent Progresses in Eco-Friendly Fabrication and Applications of Sustainable Aerogels from Various Waste Materials. *Waste and Biomass Valorization*, *13*(4), 1825–1847. DOI: 10.1007/s12649-021-01627-3 PMID: 34745389

Nita, L. E., Ghilan, A., Rusu, A. G., Neamtu, I., & Chiriac, A. P. (2020). New Trends in Bio-Based Aerogels. *Pharmaceutics*, *12*(5), 449. DOI: 10.3390/pharmaceutics12050449 PMID: 32414217

Noman, M. T., Amor, N., Ali, A., Petrik, S., Coufal, R., Adach, K., & Fijalkowski, M. (2021). Aerogels for Biomedical, Energy and Sensing Applications. *Gels (Basel, Switzerland)*, *7*(4), 264. DOI: 10.3390/gels7040264 PMID: 34940324

Nowak, D., & Jakubczyk, E. (2020). The Freeze-Drying of Foods-The Characteristic of the Process Course and the Effect of Its Parameters on the Physical Properties of Food Materials. *Foods*, *9*(10), 1488. DOI: 10.3390/foods9101488 PMID: 33080983

Paraskevopoulou, P., Chriti, D., Raptopoulos, G., & Anyfantis, G. C. (2019). Synthetic Polymer Aerogels in Particulate Form. *Materials (Basel)*, *12*(9), 1543. DOI: 10.3390/ma12091543 PMID: 31083421

Pradhan, S., Brooks, A. K., & Yadavalli, V. K. (2020). Nature-derived materials for the fabrication of functional biodevices. *Materials Today. Bio*, *7*, 100065. DOI: 10.1016/j.mtbio.2020.100065 PMID: 32613186

Sathish, S., & Shen, A. Q. (2021). Toward the Development of Rapid, Specific, and Sensitive Microfluidic Sensors: A Comprehensive Device Blueprint. *JACS Au*, *1*(11), 1815–1833. DOI: 10.1021/jacsau.1c00318 PMID: 34841402

Scott, M. J., Verhagen, W. J. C., Bieber, M. T., & Marzocca, P. (2022). A Systematic Literature Review of Predictive Maintenance for Defence Fixed-Wing Aircraft Sustainment and Operations. *Sensors (Basel)*, *22*(18), 7070. DOI: 10.3390/s22187070 PMID: 36146419

Seesaard, T., & Wongchoosuk, C. (2023). Flexible and Stretchable Pressure Sensors: From Basic Principles to State-of-the-Art Applications. *Micromachines*, *14*(8), 1638. DOI: 10.3390/mi14081638 PMID: 37630177

Sharma, S. K., Ranjani, P., Mamane, H., & Kumar, R. (2023). Preparation of graphene oxide-doped silica aerogel using supercritical method for efficient removal of emerging pollutants from wastewater. *Scientific Reports*, *13*(1), 16448. DOI: 10.1038/s41598-023-43613-w PMID: 37777623

Sheng, Z., Liu, Z., Hou, Y., Jiang, H., Li, Y., Li, G., & Zhang, X. (2023). The Rising Aerogel Fibers: Status, Challenges, and Opportunities. Advanced science (Weinheim, Baden-Wurttemberg, Germany), 10(9), e2205762. DOI: 10.1002/advs.202205762

Shirsat, M. D., & Hianik, T. (2023). Electrochemical Detection of Heavy Metal Ions Based on Nanocomposite Materials. *Journal of Composites Science*, *7*(11), 473. DOI: 10.3390/jcs7110473

Smith, S., Goodge, K., Delaney, M., Struzyk, A., Tansey, N., & Frey, M. (2020). A Comprehensive Review of the Covalent Immobilization of Biomolecules onto ElectrospunNanofibers. *Nanomaterials (Basel, Switzerland)*, *10*(11), 2142. DOI: 10.3390/nano10112142 PMID: 33121181

Song, J., Choi, S., Lim, J., & Kim, K. T. (2022). Templated synthesis of microparticles with carbonaceous skeletal structures using polymer cubosomes as templates. *RSC Advances*, *12*(14), 8429–8434. DOI: 10.1039/D2RA00832G PMID: 35424782

Sozcu, S., Venkataraman, M., Wiener, J., Tomkova, B., Militky, J., & Mahmood, A. (2023). Incorporation of Cellulose-Based Aerogels into Textile Structures. *Materials (Basel)*, *17*(1), 27. DOI: 10.3390/ma17010027 PMID: 38203881

Thakur, K., Attri, C., & Seth, A. (2021). Nanocarriers-based immobilization of enzymes for industrial application. 3 Biotech, 11(10), 427. DOI: 10.1007/s13205-021-02953-y

Ullo, S. L., & Sinha, G. R. (2020). Advances in Smart Environment Monitoring Systems Using IoT and Sensors. *Sensors (Basel)*, *20*(11), 3113. DOI: 10.3390/s20113113 PMID: 32486411

Wang, J., Petit, D., & Ren, S. (2020). Transparent thermal insulation silica aerogels. *Nanoscale Advances*, *2*(12), 5504–5515. DOI: 10.1039/D0NA00655F PMID: 36133881

Yan, X., Wu, Y., Li, R., Shi, C., Moro, R., Ma, Y., & Ma, L. (2019). High-Performance UV-Assisted NO2 Sensor Based on Chemical Vapor Deposition Graphene at Room Temperature. *ACS Omega*, *4*(10), 14179–14187. DOI: 10.1021/acsomega.9b00935 PMID: 31508539

Yu, H., Li, H., Sun, X., & Pan, L. (2023). Biomimetic Flexible Sensors and Their Applications in Human Health Detection. *Biomimetics (Basel, Switzerland)*, *8*(3), 293. DOI: 10.3390/biomimetics8030293 PMID: 37504181

Yunus, G., Singh, R., Raveendran, S., & Kuddus, M. (2023). Electrochemical biosensors in healthcare services: Bibliometric analysis and recent developments. *PeerJ*, *11*, e15566. DOI: 10.7717/peerj.15566 PMID: 37397018

Zhan, W., Chen, L., Kong, Q., Li, L., Chen, M., Jiang, J., Li, W., Shi, F., & Xu, Z. (2023). The Synthesis and Polymer-Reinforced Mechanical Properties of SiO2 Aerogels: A Review. *Molecules (Basel, Switzerland)*, *28*(14), 5534. DOI: 10.3390/molecules28145534 PMID: 37513406

Zhang, H., & Zhang, Y. (2023). Rational Design of Flexible Mechanical Force Sensors for Healthcare and Diagnosis. *Materials (Basel)*, *17*(1), 123. DOI: 10.3390/ma17010123 PMID: 38203977

Zheng, L., Zhang, S., Ying, Z., Liu, J., Zhou, Y., & Chen, F. (2020). Engineering of Aerogel-Based Biomaterials for Biomedical Applications. *International Journal of Nanomedicine*, *15*, 2363–2378. DOI: 10.2147/IJN.S238005 PMID: 32308388

Chapter 4
Application of Aerogels in Acoustic and Thermal Insulation

Abel Inobeme
Edo State University, Uzairue, Nigeria

Kingsley Erhon Enerjiofo
https://orcid.org/0000-0001-7513-4052
Glorious Vision University, Ogwa, Nigeria

ABSTRACT

Aerogels, renowned for their low density and high porosity, have garnered significant attention in the realm of acoustic and thermal insulation due to their exceptional properties. In acoustic insulation, aerogels serve as effective sound absorbers, mitigating noise transmission by dissipating sound energy through their intricate nanoporous structure. Moreover, aerogels exhibit promising potential in thermal insulation applications. With their low thermal conductivity, aerogels impede heat transfer, thereby enhancing the energy efficiency of buildings and industrial processes. This study underscores the multifaceted utility of aerogels in advancing acoustic and thermal insulation technologies, offering sustainable solutions for noise control and thermal management in diverse applications.

DOI: 10.4018/979-8-3693-5971-6.ch004

11.1 INTRODUCTION

Aerogels, often dubbed as "frozen smoke" or "solid air," represent a class of highly porous materials with remarkable properties that make them suitable for a wide range of applications, including acoustic and thermal insulation. These materials are structurally unique, typically composed of a network of interconnected nanoparticles with a high porosity, resulting in an extremely low density. This structural arrangement imparts aerogels with exceptional properties such as low thermal conductivity, high surface area, and high sound absorption capacity, making them ideal candidates for insulation purposes. In recent years, the demand for effective acoustic and thermal insulation materials has surged, driven by the need for energy-efficient buildings, noise reduction in transportation systems, and advancements in industrial processes. Traditional insulation materials, while effective to some extent, often suffer from limitations such as bulkiness, weight, and environmental concerns. Aerogels offer a compelling alternative, addressing these drawbacks while offering superior performance in acoustic and thermal insulation applications (Vagtholm et al. 2023). Aerogels are colloidal materials derived from various precursors such as silica, carbon, and metal oxides. The synthesis process involves the removal of liquid components from a gel through supercritical drying or freeze-drying, leaving behind a solid matrix with an open-porous structure. This process preserves the nanoscale architecture of the gel, resulting in a material with high porosity (typically over 90%), low density (as low as a few milligrams per cubic centimeter), and large surface area (ranging from hundreds to thousands of square meters per gram). The nanoporous structure of aerogels gives rise to their exceptional properties, including ultra-low thermal conductivity, high surface area-to-volume ratio, and superior mechanical strength despite their low density. These attributes make aerogels highly attractive for various applications, particularly in insulation, where minimizing heat transfer and sound propagation is paramount (Lázár et al. 2023). Aerogels possess several key properties that render them well-suited for acoustic and thermal insulation applications. Firstly, their low thermal conductivity, typically in the range of 0.010–0.040 W/mK, significantly reduces heat transfer through conduction, making them highly effective insulators. This property is particularly advantageous in environments where maintaining thermal comfort and energy efficiency is essential, such as buildings, refrigeration systems, and spacecraft. Secondly, aerogels exhibit excellent sound absorption characteristics due to their high porosity and tortuous pore structure. When sound waves propagate through an aerogel, they encounter numerous interfaces between the solid and gas phases, leading to multiple reflections and dissipation of sound energy. This mechanism results in effective noise reduction across a broad frequency range, making aerogels suitable for applications requiring acoustic insulation, such as automotive interiors, aircraft cabins, and architectural spaces (Jin et al. 2023). Furthermore, aerogels are lightweight and mechanically robust, offering ease of handling and installation while maintaining structural integrity over time. Their hydrophobic nature and resistance to microbial growth enhance durability and longevity in diverse operating conditions. In the subsequent sections, we delve deeper into the specific applications of aerogels in acoustic and thermal insulation, exploring their mechanisms of action, performance characteristics, case studies, and future prospects (Revin et al. 2020).

11.2 PROPERTIES OF AEROGELS RELEVANT TO ACOUSTIC AND THERMAL INSULATION

Aerogels, often dubbed as "frozen smoke" or "solid air," are remarkable materials characterized by their low density, high porosity, and unique nanoporous structure. These properties make them highly suitable for a wide range of applications, including acoustic and thermal insulation. In this section, we delve into the key properties of aerogels that are particularly relevant to their effectiveness in acoustic and thermal insulation. Low Density: One of the most striking features of aerogels is their extremely low density. Typically, aerogels have densities ranging from 0.001 g/cm^3 to 0.5 g/cm^3, making them some of the lightest solid materials known to humanity. This low density enables aerogels to provide effective insulation without adding significant weight or bulk to the insulated structure. In acoustic insulation applications, the low density of aerogels contributes to their ability to attenuate sound waves without causing structural strain or compromising the overall integrity of the system (Leventis, 2022).

High Porosity: Aerogels possess an extraordinarily high porosity, with porosity levels often exceeding 90%. This high porosity is a consequence of the aerogel synthesis process, which involves the removal of the liquid component from a gel while preserving the solid matrix's structure. The resulting nanoporous network in aerogels provides an extensive surface area, which is advantageous for both acoustic and thermal insulation applications. In acoustic insulation, the porous structure of aerogels facilitates the absorption of sound energy, thereby reducing the transmission of noise. Similarly, in thermal insulation, the high porosity of aerogels impedes the transfer of heat through the material, enhancing its insulating properties (Bakhori et al. 2023). Nanoporous Structure: Aerogels exhibit a highly porous structure at the nanometer scale, with interconnected networks of pores ranging from a few nanometers to several micrometers in size. This nanoporous structure contributes to the exceptional thermal and acoustic insulation properties of aerogels by effectively trapping air molecules within the material. In acoustic insulation, the nanoporous structure of aerogels creates multiple interfaces for sound wave interaction, leading to efficient sound absorption and attenuation. Similarly, in thermal insulation, the nanoporous structure of aerogels minimizes heat conduction by impeding the movement of air molecules through the material, thereby reducing thermal conductivity (Lou et al. 2023). Low Thermal Conductivity: Aerogels are renowned for their exceptionally low thermal conductivity, which is attributed to their nanoporous structure and low density. The low thermal conductivity of aerogels makes them highly effective thermal insulators, capable of significantly reducing heat transfer through conduction. This property is particularly advantageous in thermal insulation applications, where aerogels can be used to create thermal barriers that minimize heat loss or gain in buildings, appliances, and industrial processes. By incorporating aerogels into insulation systems, it is possible to enhance energy efficiency and reduce heating and cooling costs (Zhang et al. 2018; Mathew et al., 2024a). Hydrophobicity/Hydrophilicity: Depending on the specific composition and surface treatment, aerogels can exhibit varying degrees of hydrophobicity or hydrophilicity. This property can be advantageous in both acoustic and thermal insulation applications. In acoustic insulation, hydrophobic aerogels repel moisture, preventing the material from becoming waterlogged and maintaining its acoustic performance over time.

In thermal insulation, hydrophobic aerogels can resist moisture ingress, thereby preserving their insulating properties and preventing thermal degradation (Zhang et al. 2022). Generally, the unique combination of properties exhibited by aerogels makes them highly desirable materials for acoustic and thermal insulation applications. Their low density, high porosity, nanoporous structure, low thermal conductivity, and hydrophobic/hydrophilic nature collectively contribute to their effectiveness as insulating materials. By leveraging these properties, aerogels offer sustainable solutions for noise control, energy efficiency enhancement, and thermal management across various industries and applications (Lou et al. 2023).

11.3 APPLICATION OF AEROGELS IN ACOUSTIC INSULATION

Aerogels, with their porous and lightweight structure, offer remarkable potential for acoustic insulation across various industries (figure 1). In building construction, aerogels are utilized within walls, floors, and ceilings to mitigate noise transmission, enhancing comfort and privacy. Their high porosity enables efficient sound absorption, making them effective in reducing reverberation and dampening noise. Moreover, aerogels exhibit superior thermal insulation properties, contributing to energy efficiency in buildings. In automotive applications, aerogels find use in car doors, floors, and engine compartments to attenuate road and engine noise, enhancing the overall driving experience. Their lightweight nature is particularly advantageous in vehicles, as it helps in reducing overall weight and improving fuel efficiency. Aerogels are also deployed in aerospace industries to minimize noise pollution within aircraft interiors, ensuring a quieter and more comfortable journey for passengers. Their high strength-to-weight ratio is especially valuable in aerospace applications, contributing to weight savings without compromising structural integrity (Dong et al. 2023).

In industrial settings, aerogels are employed in machinery, equipment, and HVAC systems to provide both thermal and acoustic insulation. Their excellent insulating properties help in maintaining stable operating temperatures while minimizing noise pollution from industrial processes.

Despite their numerous benefits, challenges such as costly production methods and limited availability remain barriers to widespread adoption of aerogels in acoustic insulation applications. However, ongoing research and advancements in manufacturing techniques are expected to address these challenges, further expanding the use of aerogels in acoustic insulation across various industries (Khan et al. 2023).

Figure 1. Applications of aerogel in acoustic and insulation

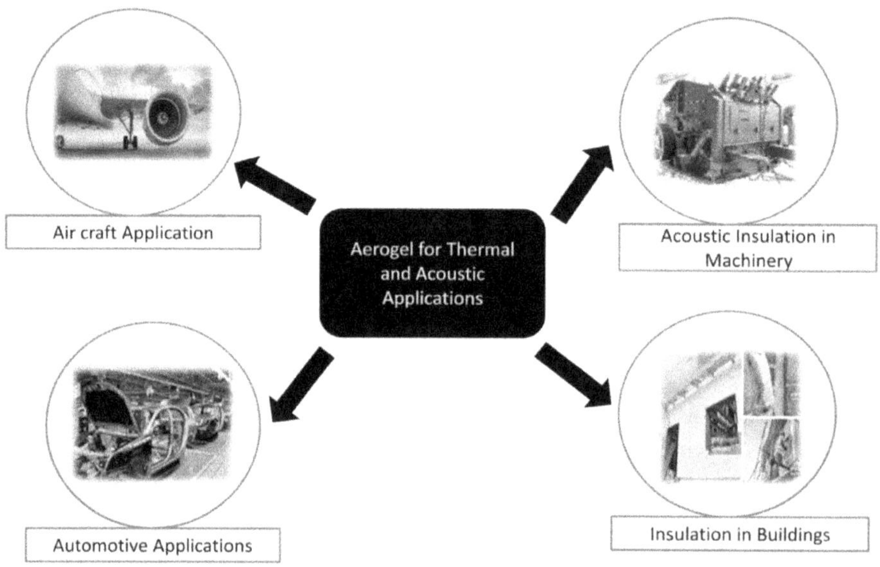

This table 11.1 provides a brief overview of the various applications of aerogels in acoustic insulation, along with their advantages and challenges.

Table 1. Application of aerogels in acoustic insulation

Application	Description	Advantages	Challenges	References
Building Insulation	Used in walls, floors, and ceilings to reduce noise	High sound absorption	Costly production	
Automotive	Used in car doors, floors, and engine compartments	Lightweight and space-saving	Limited availability	Czerwinski, 2021.
Aerospace	Used in aircraft interiors to reduce noise pollution	High strength-to-weight ratio	Fragile material	Ang and Cui, 2022.
Industrial	Used in machinery, equipment, and HVAC systems	Excellent thermal and acoustic insulator	Limited customization	An et al. 2022.

11.4 PERFORMANCE EVALUATION OF AEROGEL-BASED ACOUSTIC INSULATION

Aerogels, known for their ultra-lightweight and porous structure, have garnered significant attention in recent years for their potential applications in acoustic insulation. As industries and individuals alike seek effective solutions to mitigate noise pollution, aerogel-based acoustic insulation presents a promising avenue for achieving sound attenuation while addressing other engineering challenges such as weight reduction and thermal insulation. This paper aims to evaluate the performance of aerogel-based acoustic insulation across various applications, highlighting its effectiveness, advantages, challenges, and potential areas for improvement (Sozcu et al. 2023).

11.4.1 Advantages of Aerogel-based Acoustic Insulation

Aerogels possess several advantageous properties that make them attractive for acoustic insulation applications (figure 2). Firstly, their high porosity and interconnected nanoporous structure enable efficient absorption of sound waves across a broad frequency range. This property is crucial for achieving effective noise reduction, particularly in environments with diverse noise sources. Additionally, aerogels exhibit excellent thermal insulation properties, providing dual functionality by simultaneously reducing heat transfer and sound transmission. This dual functionality makes aerogel-based insulation ideal for applications where both acoustic and thermal comfort are paramount, such as in buildings, vehicles, and industrial machinery (Linhares et al. 2023).

Furthermore, aerogels are lightweight materials, which is advantageous for applications where weight reduction is critical, such as in aerospace and automotive industries. By incorporating aerogel-based acoustic insulation, manufacturers can achieve significant weight savings without compromising on acoustic performance. This aspect is particularly important in aerospace applications, where every gram of weight reduction translates into fuel savings and improved aircraft efficiency (Rapisarda and Meo,2022).

Figure 2. Properties of aerogel favoring its application in insulation

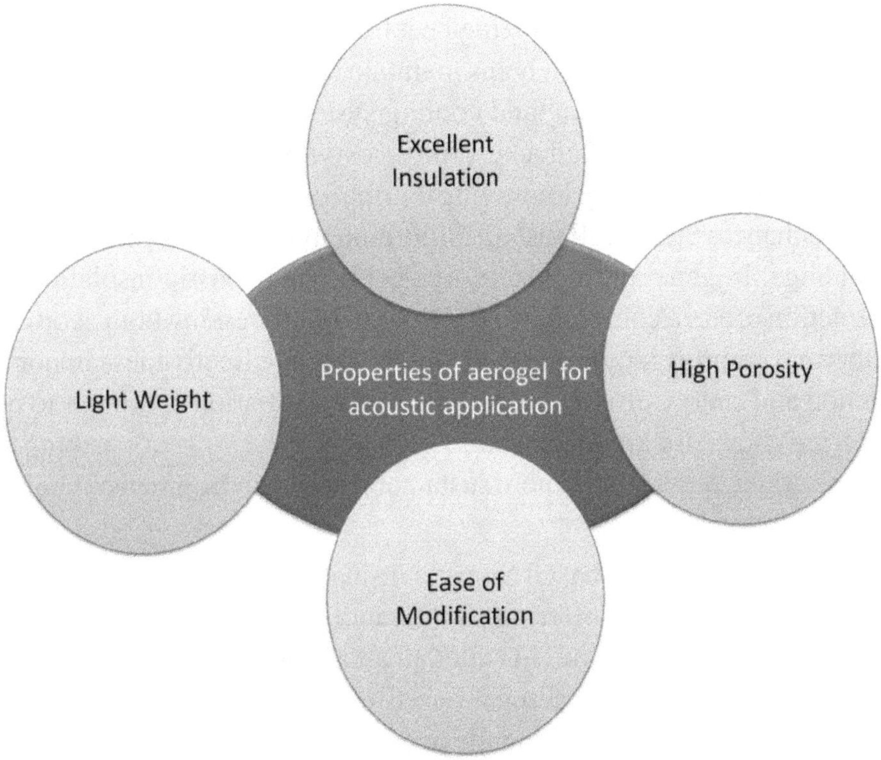

11.4.2 Performance Evaluation Across Different Applications

Building Construction: In the construction industry, aerogel-based acoustic insulation serves as a vital solution for minimizing noise transmission within buildings. Applied in walls, floors, and ceilings, aerogel-based materials effectively attenuate noise propagation between rooms and from external sources, creating quieter and more comfortable indoor environments. Performance evaluations have consistently demonstrated the ability of aerogel-based insulation to achieve high sound absorption coefficients, thereby reducing reverberation and dampening noise within buildings. This contributes to enhanced privacy and productivity for occupants, as well as improved overall quality of life. Furthermore, the thermal insulation properties of aerogels provide added benefits in terms of energy efficiency. By minimizing heat transfer through walls and ceilings, aerogel-based insulation helps maintain stable indoor temperatures, reducing the need for heating and cooling systems. This results in lower energy consumption and reduced utility costs for building owners and occupants. Additionally, by improving thermal comfort, aerogel-based insulation enhances the overall sustainability and environmental performance of buildings. In general, the use of aerogel-based acoustic insulation in construction offers a comprehensive solution for addressing both acoustic and thermal comfort requirements. With its proven effectiveness in noise reduction and energy efficiency, aerogel-based insulation continues to be a preferred choice for architects, builders, and property owners seeking to create healthier, quieter, and more sustainable built environments (Fediuk et al. 2021).

Automotive: Aerogel-based acoustic insulation is widely employed in automotive applications, offering significant benefits in noise reduction and vehicle weight reduction. Installed in areas such as car doors, floors, and engine compartments, aerogel-based materials effectively dampen road and engine noise, enhancing the overall acoustic comfort for vehicle occupants. Performance evaluations consistently show that aerogel-based insulation achieves notable reductions in noise transmission, resulting in quieter cabin environments and a more pleasant driving experience. In addition to its acoustic properties, the lightweight nature of aerogels is particularly advantageous in automotive applications. By incorporating aerogel-based insulation, vehicle manufacturers can reduce overall vehicle weight, leading to improved fuel efficiency and reduced emissions. This weight reduction contributes to meeting regulatory standards for fuel economy and environmental performance, while also enhancing the agility and handling characteristics of the vehicle. Generally, the use of aerogel-based acoustic insulation in automotive applications represents a strategic approach to improving vehicle comfort, performance, and environmental sustainability. With ongoing advancements in material science and manufacturing techniques, aerogel-based insulation is expected to play an increasingly integral role in shaping the future of quieter, more efficient, and environmentally friendly automobiles (Czerwinski, 2021).

Aerospace: In the aerospace industry, aerogel-based acoustic insulation plays a crucial role in enhancing the in-flight experience for passengers while ensuring compliance with stringent noise regulations. Utilized within aircraft interiors, aerogel-based materials effectively attenuate noise pollution, creating quieter cabin environments conducive to passenger comfort and relaxation. Performance evaluations consistently demonstrate the superior sound absorption properties of aerogel-based insulation, which contribute to minimizing noise transmission from engines, airflow, and other sources within the aircraft. One of the key advantages of aerogel-based acoustic insulation in aerospace applications is its lightweight profile. Despite its exceptional sound absorption capabilities, aerogel-based materials add minimal weight to the aircraft, supporting fuel efficiency and operational cost savings. This lightweight characteristic is particularly advantageous in aviation, where every gram of weight reduction translates into improved fuel economy and reduced emissions. Furthermore, aerogel-based insulation enables aircraft manufacturers to meet stringent regulatory requirements for noise levels within aircraft cabins. By effectively reducing noise pollution, aerogel-based materials help airlines comply with noise regulations while enhancing the overall passenger experience. This combination of superior acoustic performance, lightweight design, and regulatory compliance makes aerogel-based acoustic insulation an ideal solution for aerospace applications, contributing to quieter, more comfortable, and environmentally sustainable air travel (Jin et al. 2023).

Industrial Applications: Aerogel-based acoustic insulation is a valuable asset in industrial settings, offering multifaceted benefits. Within machinery, equipment, and HVAC systems, it serves to mitigate noise pollution, fostering quieter working environments conducive to improved focus and productivity. Performance evaluations consistently demonstrate the efficacy of aerogel-based materials in attenuating noise from various industrial processes, ensuring compliance with noise regulations and safeguarding employee health and well-being. Moreover, beyond its acoustic properties, aerogel-based insulation contributes to enhanced workplace safety by virtue of its thermal insulation capabilities. By maintaining stable operating temperatures within industrial facilities, it helps prevent overheating of machinery and equipment, reducing the risk of accidents and equipment failures. This dual functionality of aerogel-based insulation not only promotes a safer working environment but also supports operational efficiency and longevity of industrial assets. Overall, the adoption of aerogel-based acoustic insulation in industrial settings represents a strategic investment in employee welfare, regulatory compliance, and operational excellence. As industries continue to prioritize safety, efficiency, and sustainability, the role of aerogel-based insulation in creating optimized industrial environments is poised to expand (Buratti et al. 2014; Mathew et al., 2024b).

While aerogel-based acoustic insulation offers numerous advantages, several challenges must be addressed to further enhance its performance and widespread adoption. One significant challenge is the high cost associated with aerogel production, which limits its affordability for certain applications. Research efforts focused on developing cost-effective manufacturing techniques and scaling up production could help address this challenge and make aerogel-based insulation more accessible (Bakhori et al. 2023).

Another challenge is the fragility of aerogel materials, which can be prone to damage during handling and installation. Improvements in material durability and robustness are essential to ensure long-term performance and reliability in real-world applications. Additionally, enhancing the mechanical properties of aerogels without compromising their acoustic performance could expand their utility in high-stress environments, such as in automotive and industrial applications. Furthermore, increasing the availability and variety of aerogel-based insulation products tailored to specific application requirements could accelerate their adoption across different industries. Customization options such as varying densities, thicknesses, and surface treatments can optimize acoustic performance while meeting the unique needs of each application (Chen et al. 2023).

Aerogel-based acoustic insulation holds great promise for addressing noise pollution challenges across various industries. Performance evaluations have demonstrated its effectiveness in attenuating sound transmission, improving acoustic comfort, and enhancing energy efficiency. Despite challenges such as high production costs and material fragility, ongoing research and development efforts are poised to overcome these obstacles and unlock the full potential of aerogel-based insulation. With continued innovation and investment, aerogel-based acoustic insulation is expected to play an increasingly important role in creating quieter, more comfortable, and environmentally sustainable environments (Sozcu et al. 2023)

11.5 APPLICATION OF AEROGELS IN THERMAL INSULATION

Aerogels, known for their ultralightweight and highly porous structure, have emerged as promising materials for thermal insulation applications across various industries. With superior thermal conductivity properties compared to traditional insulation materials, aerogels offer significant advantages in terms of energy efficiency, space-saving, and sustainability. This paper explores the diverse applications of aerogels in thermal insulation, highlighting their unique properties, advantages, challenges, and potential areas for further development (Khan et al. 2023).

11.5.1 Advantages of Aerogels in Thermal Insulation

Aerogels possess several advantageous properties that make them highly effective for thermal insulation. Firstly, their extremely low thermal conductivity, typically in the range of 0.010 to 0.030 W/mK, enables superior heat insulation compared to conventional materials such as fiberglass or foam. This property allows aerogels to provide high levels of insulation with minimal thickness, making them ideal for applications where space constraints are a concern, such as in buildings, appliances, and aerospace components (Thie et al. 2023).

Moreover, aerogels exhibit excellent thermal stability, retaining their insulation properties over a wide temperature range, from cryogenic temperatures to elevated temperatures exceeding 400°C. This thermal stability makes aerogels suitable for use in extreme environments, including industrial processes, automotive engines, and space exploration missions. Another advantage of aerogels is their hydrophobic nature, which prevents moisture absorption and maintains insulation performance even in humid or wet conditions. This property is particularly important for outdoor applications, where exposure to rain, snow, or high humidity levels can compromise the effectiveness of traditional insulation materials (Arshad et al. 2023).

Furthermore, aerogels are lightweight and chemically inert, making them easy to handle and compatible with a wide range of construction materials, including plastics, metals, and composites. Their lightweight nature is advantageous in applications where weight reduction is critical, such as in aircraft, vehicles, and portable electronic devices (Nita et al. 2020).

11.5.2 Applications of Aerogels in Thermal Insulation

Building Construction: In the construction industry, aerogels play a vital role in enhancing the thermal performance of buildings. Aerogel-based insulation materials are utilized in walls, roofs, floors, and windows to minimize heat transfer and enhance energy efficiency. With their low thermal conductivity and high insulation efficiency, aerogels enable builders to meet rigorous energy codes and obtain sustainable building certifications such as LEED (Leadership in Energy and Environmental Design). By incorporating aerogels into building envelopes, construction projects can achieve improved thermal comfort for occupants, reduced energy consumption, and lower environmental impact, contributing to a more sustainable built environment. Overall, the use of aerogels in construction underscores their importance in promoting energy efficiency and environmental sustainability in building design and construction practices (Ganobjak et al. 2023).

Appliances: Aerogels play a significant role in improving the thermal insulation and energy efficiency of household appliances such as refrigerators, freezers, and ovens. By incorporating aerogel-based insulation into the walls and doors of these appliances, manufacturers can achieve several benefits. Firstly, aerogel-based insulation effectively minimizes heat transfer, helping to maintain stable internal temperatures within refrigerators, freezers, and ovens. This results in improved temperature control, ensuring that food remains fresh and properly preserved while also optimizing cooking performance. Secondly, the use of aerogel insulation leads to reduced energy consumption by these appliances. By minimizing heat loss or gain, aerogels enable refrigerators, freezers, and ovens to operate more efficiently, thereby lowering electricity usage and decreasing utility costs for consumers. Furthermore, aerogel-based insulation can prolong the lifespan of household appliances. By reducing thermal stress on internal components and improving overall temperature stability, aerogels help to mitigate wear and tear, extending the durability and longevity of refrigerators, freezers, and ovens. Additionally, the integration of aerogels into household appliances offers a practical and effective solution for enhancing thermal performance, energy efficiency, and product longevity, thereby benefiting both consumers and the environment (Thie et al. 2023).

Transportation: In the automotive and aerospace industries, aerogels serve as effective thermal insulation materials for various vehicle components, including engine compartments, exhaust systems, and cabin interiors. Aerogel-based insulation minimizes heat transfer, reducing thermal stress on engine components and improving overall engine performance and longevity. Additionally, in cabin interiors, aerogel insulation helps maintain stable temperatures, enhancing passenger comfort during both automotive and aerospace travel. The lightweight nature of aerogels is particularly advantageous in these industries, as it contributes to weight reduction, fuel efficiency, and improved vehicle performance. Overall, the utilization of aerogels in automotive and aerospace applications underscores their importance in enhancing thermal management, optimizing vehicle performance, and improving the overall passenger experience (Jin et al.2023).

Industrial Applications: Aerogels play a crucial role in industrial applications, offering exceptional thermal insulation properties that enhance energy efficiency and operational reliability. In environments such as oil and gas pipelines, chemical reactors, and cryogenic storage tanks, aerogel-based insulation serves to minimize heat loss, preventing energy wastage and ensuring process efficiency. By reducing thermal conductivity, aerogels help maintain consistent temperatures, preventing fluctuations that could affect production processes or product quality. Additionally, aerogel-based insulation helps prevent condensation in these industrial environments. By creating a barrier that inhibits heat transfer between the interior and exterior surfaces, aerogels minimize temperature differentials that can lead to moisture buildup and corrosion. This not only protects equipment and infrastructure from damage but also improves safety and longevity. Furthermore, in cryogenic applications, aerogel-based insulation is particularly valuable for its ability to maintain low temperatures without excessive heat transfer. This ensures the integrity of cryogenic storage tanks and equipment while minimizing energy consumption and operational costs. Furthermore, the use of aerogels in industrial applications contributes to improved energy efficiency, process stability, and operational reliability, making them indispensable for a wide range of industrial processes and infrastructure (Noman et al. 2021).

While aerogels offer significant advantages for thermal insulation, several challenges remain to be addressed to realize their full potential. One challenge is the high cost of aerogel production, which limits their widespread adoption, particularly in large-scale applications such as building construction and industrial insulation. Research efforts focused on developing cost-effective production methods and scalable manufacturing processes are needed to overcome this challenge. Another challenge is the brittleness and fragility of aerogel materials, which can pose difficulties during handling, transportation, and installation. Improvements in material processing and reinforcement techniques are required to enhance the mechanical strength and durability of aerogels, making them more suitable for practical applications in harsh or dynamic environments. Furthermore, while aerogels exhibit excellent thermal insulation properties, their performance may be affected by factors such as compression, moisture exposure, and aging. Continued research is needed to understand and mitigate these effects, ensuring long-term stability and reliability of aerogel-based insulation systems (Shan et al. 2024).

Aerogels offer a compelling solution for thermal insulation across a wide range of applications, from building construction to industrial processes and transportation systems. With their exceptional thermal conductivity, stability, and hydrophobic properties, aerogels enable improved energy efficiency, space-saving designs, and enhanced environmental sustainability. While challenges such as high production costs and material fragility remain to be addressed, ongoing research and development efforts are expected to further expand the use of aerogels in thermal insulation, driving innovation and performance improvements in the field (Liu et al. 2020).

11.6 COMBINED ACOUSTIC AND THERMAL INSULATION SYSTEMS INCORPORATING AEROGELS

Combined acoustic and thermal insulation systems incorporating aerogels offer multifunctional solutions across industries (**Table 2**). In building envelopes, aerogels installed in walls, roofs, and windows effectively reduce both noise transmission and heat transfer, improving thermal comfort and energy efficiency. Automotive applications benefit from aerogel-based insulation in vehicle interiors, dampening noise and regulating interior temperatures simultaneously, enhancing acoustic comfort and reducing energy consumption. Aerospace industries utilize aerogels in aircraft cabins to minimize noise pollution and thermal fluctuations, ensuring passenger comfort and energy efficiency. In industrial settings, aerogel-based insulation in machinery and HVAC systems mitigates noise pollution and minimizes heat transfer, enhancing workplace safety and energy efficiency. Challenges include the cost of aerogel materials, integration with existing structures, space constraints in vehicles, weight restrictions in aircraft, and compatibility with industrial equipment and materials. Nonetheless, combined acoustic and thermal insulation systems incorporating aerogels offer significant advantages in improving comfort, efficiency, and environmental sustainability (Zhou, 2021).

Table 2. Combined acoustic and thermal insulation systems incorporating aerogels

Application	Description	Advantages	Challenges	References
Building Envelopes	Used in walls, roofs, and windows to reduce bothnoisetransmission and heat transfer	Improved thermal comfort and energy efficiency Reduced noise pollution within buildings	Cost of aerogel materials Integration with existing building materials	Kumar et al. 2021.
Automotive	Implemented in vehicle interiors, particularly in doors and floors, to dampen noise and regulate interior temperatures	Enhanced acoustic comfort and thermal insulation Reduced energy consumption and improved passenger comfort	Space constraints within vehicle components Compatibility with automotive materials	Ahmed et al. 2022.
Aerospace	Deployed in aircraft cabins to minimize both noise pollution and thermal fluctuations	Enhanced passenger comfort and energy efficiency Improved thermal insulation in extreme temperatures	Weight restrictions in aircraft design Compatibility with aircraft materials and regulations	Bagshaw & Illig, 2019.
Industrial	Utilized in machinery, equipment, and HVAC systems to mitigate noise pollution and minimize heat transfer	Reduced noise levels and thermal conductivity Improved workplace safety and energy efficiency	Compatibility with industrial equipment and materials Cost-effectiveness in industrial applications	Akadiri et al. 2012.

11.7 CONCLUSION

The application of aerogels in combined acoustic and thermal insulation systems offers a versatile and efficient solution across various industries. Aerogels provide exceptional thermal insulation properties, effectively reducing heat transfer and enhancing energy efficiency in buildings, vehicles, aircraft, and industrial equipment. Additionally, their high porosity and interconnected nanoporous structure enable efficient absorption of sound waves, effectively attenuating noise pollution and improving acoustic comfort. By incorporating aerogels into insulation systems, builders, manufacturers, and engineers can achieve significant benefits, including improved thermal comfort, reduced energy consumption, enhanced acoustic performance, and compliance with regulatory standards. However, challenges such as the cost of aerogel materials, integration with existing structures, and compatibility with industry-specific requirements remain to be addressed. Nonetheless, ongoing research and development efforts focused on optimizing aerogel production techniques, enhancing material durability, and improving integration methods are expected to overcome these challenges, further expanding the use of aerogels in combined acoustic and thermal insulation applications. With their unique combination of properties, aerogels continue to demonstrate their potential to revolutionize insulation technology, paving the way for more sustainable, comfortable, and efficient built environments.

REFERENCES

Ahmed, A. T. M. F., Islam, M. Z., Mahmud, M. S., Sarker, M. E., & Islam, M. R. (2022). Hemp as a potential raw material toward a sustainable world: A review. *Heliyon, 8*(1), e08753. DOI: 10.1016/j.heliyon.2022.e08753 PMID: 35146149

An, L., Guo, Z., Li, Z., Fu, Y., Hu, Y., Huang, Y., Yao, F., Zhou, C., & Ren, S. (2022). Tailoring thermal insulation architectures from additive manufacturing. *Nature Communications, 13*(1), 4309. DOI: 10.1038/s41467-022-32027-3 PMID: 35879371

Ang, L. Y. L., & Cui, F. (2022). Remote work: Aircraft noise implications, prediction, and management in the built environment. Applied acoustics. Acoustique applique. *Applied Acoustics, 198*, 108978. DOI: 10.1016/j.apacoust.2022.108978 PMID: 36034578

Arshad, A., Roy, A., Mallick, T. K., & Tahir, A. A. (2023). Shape-Stabilized PEGylated Silica Aerogel-Composite as an Energy Saving Building Material. *Industrial & Engineering Chemistry Research, 62*(47), 20236–20250. DOI: 10.1021/acs.iecr.3c02373 PMID: 38045733

Bagshaw, M., &Illig, P. (2019). The Aircraft Cabin Environment. Travel Medicine, 429–436. DOI: 10.1016/B978-0-323-54696-6.00047-1

Bakhori, N. M., Ismail, Z., Hassan, M. Z., & Dolah, R. (2023). Emerging Trends in Nanotechnology: Aerogel-Based Materials for Biomedical Applications. *Nanomaterials (Basel, Switzerland), 13*(6), 1063. DOI: 10.3390/nano13061063 PMID: 36985957

Buratti, C., Moretti, E., Belloni, E., & Agosti, F. (2014). Development of Innovative Aerogel Based Plasters: Preliminary Thermal and Acoustic Performance Evaluation. *Sustainability (Basel), 6*(9), 5839–5852. DOI: 10.3390/su6095839

Chen, Z., Fu, X., Liu, R., Song, Y., & Yin, X. (2023). Fabrication, Performance, and Potential Applications of MXene Composite Aerogels. *Nanomaterials (Basel, Switzerland), 13*(14), 2048. DOI: 10.3390/nano13142048 PMID: 37513059

Czerwinski, F. (2021). Current Trends in Automotive Lightweighting Strategies and Materials. *Materials (Basel), 14*(21), 6631. DOI: 10.3390/ma14216631 PMID: 34772154

Czerwinski, F. (2021). Current Trends in Automotive Lightweighting Strategies and Materials. *Materials (Basel), 14*(21), 6631. DOI: 10.3390/ma14216631 PMID: 34772154

Dong, Y., Kong, J., Mousavi, S., Rismanchi, B., & Yap, P.-S. (2023). Wall Insulation Materials in Different Climate Zones: A Review on Challenges and Opportunities of Available Alternatives. *Thermo, 3*(1), 38–65. DOI: 10.3390/thermo3010003

Fediuk, R., Amran, M., Vatin, N., Vasilev, Y., Lesovik, V., & Ozbakkaloglu, T. (2021). Acoustic Properties of Innovative Concretes: A Review. *Materials (Basel), 14*(2), 398. DOI: 10.3390/ma14020398 PMID: 33466943

Ganobjak, M., Brunner, S., Hofmann, J., Klar, V., Ledermann, M., Herzog, V., Kämpfen, B., Kilian, R., Wehdorn, M., & Wernery, J. (2023). Current Trends in Aerogel Use in Heritage Buildings: Case Studies from the Aerogel Architecture Award 2021. *Gels (Basel, Switzerland), 9*(10), 814. DOI: 10.3390/gels9100814 PMID: 37888387

Jin, R., Zhou, Z., Liu, J., Shi, B., Zhou, N., Wang, X., Jia, X., Guo, D., & Xu, B. (2023). Aerogels for Thermal Protection and Their Application in Aerospace. *Gels (Basel, Switzerland), 9*(8), 606. DOI: 10.3390/gels9080606 PMID: 37623061

Khan, N. R., Sharmin, T., & Bin Rashid, A. (2023). Exploring the Versatility of Aerogels: Broad Applications in Biomedical Engineering, Astronautics, Energy Storage, Biosensing, and Current Progress. *Heliyon, 10*(1), e23102. DOI: 10.1016/j.heliyon.2023.e23102 PMID: 38163169

Kumar, D., Alam, M., & Sanjayan, J. G. (2021). Retrofitting Building Envelope Using Phase Change Materials and Aerogel Render for Adaptation to Extreme Heatwave: A Multi-Objective Analysis Considering Heat Stress, Energy, Environment, and Cost. *Sustainability (Basel), 13*(19), 10716. DOI: 10.3390/su131910716

Lázár, I., Čelko, L., & Menelaou, M. (2023). Aerogel-Based Materials in Bone and Cartilage Tissue Engineering-A Review with Future Implications. *Gels (Basel, Switzerland), 9*(9), 746. DOI: 10.3390/gels9090746 PMID: 37754427

Leventis, N. (2022). Polyurea Aerogels: Synthesis, Material Properties, and Applications. *Polymers, 14*(5), 969. DOI: 10.3390/polym14050969 PMID: 35267798

Linhares, T., Carneiro, V. H., Pessoa de Amorim, M. T., & Durães, L. (2023). A Comparative Thermoacoustic Insulation Study of Silica Aerogels Reinforced with Reclaimed Textile Fibres: Cotton, Polyester and Wool. *Gels (Basel, Switzerland), 9*(7), 548. DOI: 10.3390/gels9070548 PMID: 37504426

Liu, C., Wu, S., Yang, Z., Sun, H., Zhu, Z., Liang, W., & Li, A. (2020). Mechanically Robust and Flame-Retardant Silicon Aerogel Elastomers for Thermal Insulation and Efficient Solar Steam Generation. *ACS Omega, 5*(15), 8638–8646. DOI: 10.1021/acsomega.0c00086 PMID: 32337427

Lou, F., Dong, S., Zhu, K., Chen, X., & Ma, Y. (2023). Thermal Insulation Performance of Aerogel Nano-Porous Materials: Characterization and Test Methods. *Gels (Basel, Switzerland), 9*(3), 220. DOI: 10.3390/gels9030220 PMID: 36975669

Mathew, J. T., Inobeme, A., Musah, M., Azeh, Y., Abdullahi, A., Shaba, E. Y., Salihu, A. M., Muhammad, E. B., Josiah, J. G., Jibrin, N. A., Ismail, H., Muhammad, A. I., Maurice, J., Mamman, A., & Ndamitso, M. M. (2024). a. A Critical Review of Green Approach on Wastewater Treatment Strategies. *Journal of Applied Science & Environmental Management, 28*(2), 363–391. DOI: 10.4314/jasem.v28i2.9

Mathew, J. T., Inobeme, A., Musah, M., Azeh, Y., Abdullahi, A., Shaba, E. Y., Salihu, A. M., Muhammad, E. B., Josiah, J. G., Jibrin, N. A., Ismail, H., Muhammad, A. I., Maurice, J., Mamman, A., & Ndamitso, M. M. (2024). b. A Critical Review of Green Approach on Wastewater Treatment Strategies. *Journal of Applied Science & Environmental Management, 28*(2), 363–391. DOI: 10.4314/jasem.v28i2.9

Nita, L. E., Ghilan, A., Rusu, A. G., Neamtu, I., & Chiriac, A. P. (2020). New Trends in Bio-Based Aerogels. *Pharmaceutics*, *12*(5), 449. DOI: 10.3390/pharmaceutics12050449 PMID: 32414217

Noman, M. T., Amor, N., Ali, A., Petrik, S., Coufal, R., Adach, K., & Fijalkowski, M. (2021). Aerogels for Biomedical, Energy and Sensing Applications. *Gels (Basel, Switzerland)*, *7*(4), 264. DOI: 10.3390/gels7040264 PMID: 34940324

Rapisarda, M., & Meo, M. (2022). Multifunctional Thermal, Acoustic, and Piezoresistive Properties of In Situ-Modified Composite Aerogels with Graphene Oxide as the Main Phase. *ACS Applied Materials & Interfaces*, *14*(38), 43646–43655. DOI: 10.1021/acsami.2c08042 PMID: 36121008

Revin, V. V., Nazarova, N. B., Tsareva, E. E., Liyaskina, E. V., Revin, V. D., & Pestov, N. A. (2020). Production of Bacterial Cellulose Aerogels With Improved Physico-Mechanical Properties and Antibacterial Effect. *Frontiers in Bioengineering and Biotechnology*, *8*, 603407. DOI: 10.3389/fbioe.2020.603407 PMID: 33344435

Shan, J., Shan, Y., Zou, C., Hong, Y., Liu, J., & Guo, X. (2024). Cost-Effective Preparation of Hydrophobic and Thermal-Insulating Silica Aerogels. *Nanomaterials (Basel, Switzerland)*, *14*(1), 119. DOI: 10.3390/nano14010119 PMID: 38202574

Sozcu, S., Venkataraman, M., Wiener, J., Tomkova, B., Militky, J., & Mahmood, A. (2023). Incorporation of Cellulose-Based Aerogels into Textile Structures. *Materials (Basel)*, *17*(1), 27. DOI: 10.3390/ma17010027 PMID: 38203881

Thie, C., Quallen, S., Ibrahim, A., Xing, T., & Johnson, B. (2023). Study of Energy Saving Using Silica Aerogel Insulation in a Residential Building. *Gels (Basel, Switzerland)*, *9*(2), 86. DOI: 10.3390/gels9020086 PMID: 36826255

Vagtholm, R., Matteo, A., Vand, B., & Tupenaite, L. (2023). Evolution and Current State of Building Materials, Construction Methods, and Building Regulations in the U.K.: Implications for Sustainable Building Practices. *Buildings (Basel, Switzerland)*, *13*(6), 1480. DOI: 10.3390/buildings13061480

Zhang, H., Zhang, C., Ji, W., Wang, X., Li, Y., & Tao, W. (2018). Experimental Characterization of the Thermal Conductivity and Microstructure of Opacifier-Fiber-Aerogel Composite. *Molecules (Basel, Switzerland)*, *23*(9), 2198. DOI: 10.3390/molecules23092198 PMID: 30200271

Zhang, S., Yang, Z., Huang, X., Wang, J., Xiao, Y., He, J., Feng, J., Xiong, S., & Li, Z. (2022). Hydrophobic Cellulose Acetate Aerogels for Thermal Insulation. *Gels (Basel, Switzerland)*, *8*(10), 671. DOI: 10.3390/gels8100671 PMID: 36286172

Zhou, Y. (2021). Artificial neural network-based smart aerogel glazing in low-energy buildings: A state-of-the-art review. *iScience*, *24*(12), 103420. DOI: 10.1016/j.isci.2021.103420 PMID: 34849473

Chapter 5
Biomedical Applications of Aerogel:
Biosensing, Imaging, and Bone Regeneration

Stanley Okonkwo
https://orcid.org/0000-0001-5517-3193
Osaka Kyoiku University, Osaka, Japan

Abel Inobeme
Edo State University, Uzairue, Nigeria

Daniel Apeh
https://orcid.org/0000-0003-0679-2996
Confluence University of Science and Technology, Osara, Nigeria

Jonathan Inobeme
Ahmadu Bello University, Zaria, Nigeria

ABSTRACT

Aerogels have garnered significant attention in biomedical applications due to their unique properties. This review explores three key areas: biosensing, imaging, and bone regeneration. In biosensing, aerogels serve as excellent platforms for immobilizing biomolecules, enabling sensitive

DOI: 10.4018/979-8-3693-5971-6.ch005

detection of analytes through various mechanisms such as fluorescence or surface plasmon resonance. In imaging, their low density and high surface area make aerogels promising candidates for contrast agents and drug delivery vehicles in magnetic resonance imaging and computed tomography. Furthermore, their biocompatibility and tunable porosity render aerogels suitable scaffolds for bone regeneration, promoting cell adhesion, proliferation, and mineralization. This abstract underscores the multifaceted roles aerogels play in addressing diverse biomedical challenges, paving the way for future breakthroughs in healthcare.

8.1 INTRODUCTION

In recent years, aerogels have emerged as versatile materials with tremendous potential in various fields, including biomedicine. Their unique combination of properties, such as high surface area, porosity, biocompatibility, and tunable characteristics, makes them particularly well-suited for a wide range of biomedical applications. In this introduction, we will explore the diverse roles that aerogels play in biosensing, imaging, and bone regeneration within the realm of biomedicine (Bakhori et al. 2023).

Biosensing involves the detection and quantification of biological analytes, such as proteins, nucleic acids, and small molecules, for applications ranging from medical diagnostics to environmental monitoring. Aerogels serve as excellent platforms for biosensing due to their ability to immobilize biomolecules and enhance signal detection. Functionalization of aerogel surfaces with reactive groups enables the covalent attachment or physical adsorption of biomolecules, ensuring their stable anchoring onto the scaffold. This immobilization facilitates specific recognition and binding of target analytes, forming the basis of various biosensing platforms (Almeida et al. 2024).

Furthermore, the porous structure of aerogels allows for efficient diffusion of analytes, enabling rapid interaction with immobilized biomolecules and enhancing the sensitivity of biosensors. By entrapping signaling molecules or nanoparticles within the aerogel matrix, signal amplification can be achieved, further improving detection limits and sensitivity. Additionally, aerogels can act as optical or electrochemical transducers, converting biological interactions into measurable signals, thus expanding the capabilities of biosensing platforms (Noman et al. 2021).

Aerogels also play a pivotal role in imaging applications within biomedicine. The high surface area-to-volume ratio and tunable properties of aerogels make them ideal candidates for various imaging modalities, including magnetic resonance imaging (MRI), computed tomography (CT), optical imaging, and nuclear imaging techniques like positron emission tomography (PET) scans. Aerogels can be engineered to serve as contrast agents, drug delivery vehicles, or imaging probes, enhancing imaging contrast, specificity, and sensitivity (Bakhori et al., 2023).

Moreover, aerogels can be loaded with imaging agents or contrast-enhancing materials, enabling targeted delivery to specific sites within the body and controlled release for improved imaging outcomes. The stability and biocompatibility of aerogels ensure their compatibility with biological systems, minimizing adverse effects and maximizing imaging efficacy. Additionally, aerogels with tailored optical properties can be used in optical imaging techniques for biological samples, offering high transparency and light scattering properties for enhanced visualization (Nita et al. 2020).

In the realm of regenerative medicine, aerogels hold promise for bone regeneration applications. Bone defects resulting from trauma, disease, or surgery represent a significant clinical challenge, necessitating the development of advanced biomaterials for tissue repair and regeneration. Aerogels offer several advantages as scaffolds for bone regeneration, including their high porosity, biocompatibility, and ability to mimic the natural extracellular matrix of bone tissue (Lázár et al. 2023).

By serving as three-dimensional scaffolds, aerogels provide a supportive environment for the delivery of bone-forming cells, such as mesenchymal stem cells (MSCs), to the defect site. The porous structure of aerogels facilitates nutrient and oxygen diffusion to encapsulated cells while providing mechanical support. Additionally, aerogels can be engineered to exhibit osteoconductive and osteoinductive properties, promoting the formation and differentiation of bone cells for efficient tissue regeneration (Afewerki et al. 2018; Inobeme et al., 2024).

Furthermore, aerogels can be loaded with growth factors, cytokines, or other bioactive molecules that stimulate bone regeneration processes, enhancing healing outcomes. Controlled release of these factors from the aerogel matrix ensures sustained presence and localized delivery, promoting tissue regeneration directly at the site of injury. Overall, aerogel-based scaffolds hold immense potential for addressing critical needs in bone regeneration and orthopedic applications, offering innovative solutions for tissue repair and functional restoration (Zheng et al. 2020).

In addition, aerogels represent a class of multifunctional materials with diverse applications in biomedicine, spanning biosensing, imaging, and bone regeneration (figure 1). Their unique properties and tunable characteristics make them invaluable tools for addressing various challenges in healthcare, from disease diagnosis and treatment to tissue engineering and regenerative medicine. As research and development in aerogel-based technologies continue to advance, we anticipate further breakthroughs and innovations that will revolutionize biomedical science and improve patient care outcomes (Fijalkowski et al. 2023).

Figure 1. Schematic representation of chapter

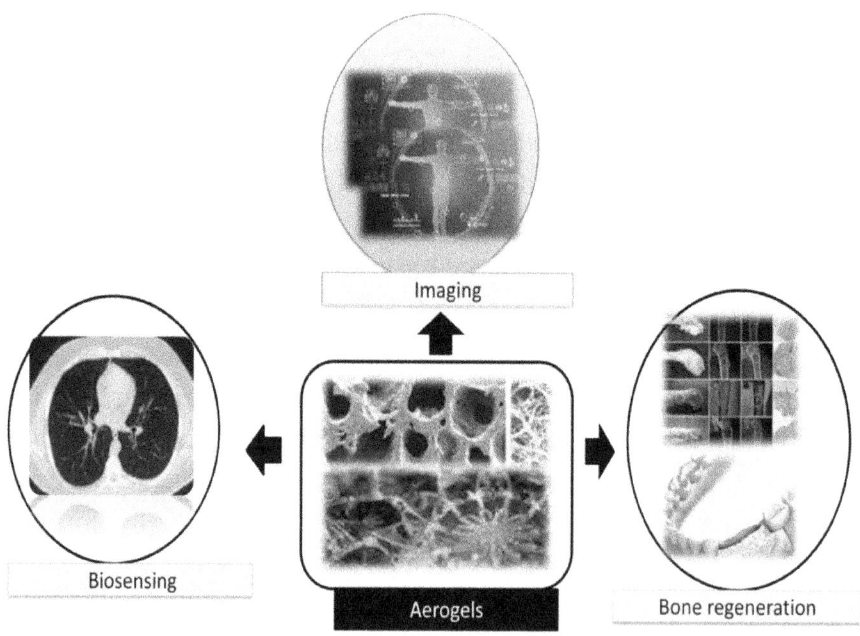

8.2 AEROGELS: PROPERTIES AND CHARACTERISTICS

Aerogels are remarkable materials known for their unique properties and characteristics (**Table 1**). Firstly, they possess an incredibly high porosity, featuring a porous structure with pore sizes ranging from nanometers to micrometers. This high porosity results in a remarkably high surface area-to-volume ratio, facilitating efficient diffusion of gases and liquids, making them ideal for applications such as adsorption and catalysis. Secondly, aerogels exhibit exceptionally high surface areas, typically ranging from several hundred to over 1000 square meters per gram. This extensive surface area allows for increased interaction with surrounding molecules, enabling applications in areas such as adsorption, catalysis, and sensing (Horvat et al. 2022; Mathew et al., 2023a).

Thirdly, aerogels are renowned for their low density, often less than 0.1 g/cm^3. This low density makes them exceptionally lightweight materials, suitable for applications where weight is a concern, such as aerospace and lightweight structural components. Finally, aerogels possess outstanding thermal insulation properties, with low thermal conductivity compared to conventional materials. This thermal insulation capability makes them highly effective for use in extreme temperature environments, providing superior thermal insulation in applications such as construction, energy-efficient windows, and thermal clothing. Overall, the unique combination of properties and characteristics of aerogels makes them valuable materials with diverse applications in various fields (Khan et al. 2023).

Table 1. Aerogels: Properties and characteristics

Property/ Characteristic	Description	Benefits	Examples	References
Porosity	Aerogels possess a highly porous structure, with pore sizes ranging from nanometers to micrometers.	Facilitates high surface area-to-volume ratio, enabling efficient diffusion of gases and liquids	Aerogels used as adsorbents for gas storage	Zhang et al. 2022; Mathew et al., 2023b
Surface Area	Aerogels exhibit exceptionally high surface area due to their porous nature, typically ranging from several hundred to over 1000 square meters per gram.	Allows for increased interaction with surrounding molecules, facilitating adsorption, catalysis, and sensing	Aerogels utilized as catalyst supports	Sheng et al. 2023.
Low Density	Aerogels are renowned for their low density, often less than 0.1 g/cm^3.	Enables lightweight materials with excellent thermal insulation properties	Aerogel-based insulation materials used in spacecraft	Khan et al. 2023.
Thermal Insulation	Aerogels possess outstanding thermal insulation properties, with low thermal conductivity compared to conventional materials.	Provides effective thermal insulation in extreme temperature environments	Aerogel blankets used in construction for energy efficiency	Thie et al. 2023.

8.3 BIOSENSING APPLICATIONS OF AEROGEL

Aerogels, with their unique properties such as high surface area, tunable porosity, and biocompatibility, have garnered significant interest in biosensing applications. Biosensing involves the detection and quantification of biological analytes, ranging from small molecules to biomolecules like proteins and nucleic acids. Aerogels serve as excellent platforms for biosensing due to their ability to immobilize biomolecules, enhance signal detection, and provide a stable environment for sensing elements (Bakhori et al. 2023). The biosensing application of aerogel has numerous merits (figure 2).

Figure 2. Merits of biosensing applications of aerogels

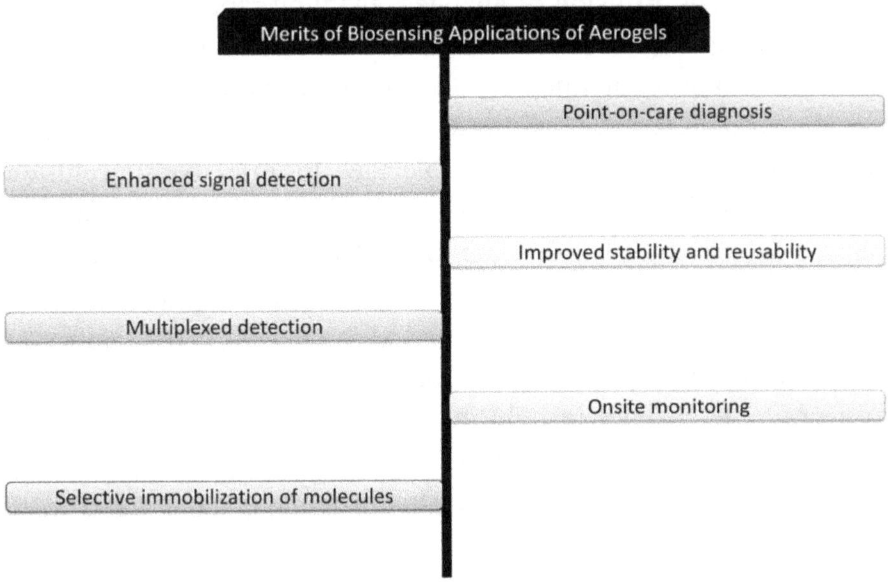

Immobilization of Biomolecules: Aerogels possess a high surface area-to-volume ratio, rendering them excellent candidates for immobilizing biomolecules like antibodies, enzymes, and DNA probes. Through surface functionalization with reactive groups, aerogels enable the covalent attachment or physical adsorption of biomolecules, ensuring their stable anchoring onto the scaffold. This immobilization provides a platform for specific recognition and binding of target analytes, serving as the foundation for various biosensing applications. Covalent attachment offers robust and permanent binding, while physical adsorption allows for reversible interactions. By tailoring the surface chemistry of aerogels, researchers can optimize the orientation and density of immobilized biomolecules, enhancing sensor performance in terms of sensitivity, selectivity, and stability. Aerogel-based biosensing platforms find applications in medical diagnostics, environmental monitoring, and food safety, offering rapid and accurate detection of analytes ranging from pathogens to biomarkers. Overall, the high surface area and versatile surface chemistry of aerogels make them indispensable tools in the development of advanced biosensors for diverse analytical challenges (Smith et al. 2020; Mathew et al., 2024a).

Enhanced Signal Detection: The porous structure of aerogels facilitates the efficient diffusion of analytes, enabling rapid interaction with immobilized biomolecules and enhancing the sensitivity of biosensors. Moreover, this porous architecture allows for the entrapment of signaling molecules or nanoparticles within the aerogel matrix, leading to signal amplification. By entrapping these molecules, the effective concentration of signaling species near the sensing elements is increased, resulting in enhanced detection signals and improved detection limits. Furthermore, aerogels can serve as versatile transducers in biosensing applications. In optical biosensors, aerogels with tailored optical properties can act as optical transducers, facilitating the conversion of biological interactions into measurable changes in light intensity, wavelength, or polarization. Similarly, in electrochemical biosensors, aerogels can function as electrochemical transducers, translating biochemical reactions into measurable electrical signals. Overall, the combination of efficient analyte diffusion, signal amplification, and transducer capabilities makes aerogels valuable components in biosensing platforms, enabling sensitive, selective, and rapid detection of biological analytes for various applications in healthcare, environmental monitoring, and beyond (Le & Yoon, 2024).

Stability and Reusability: Aerogels offer a stable and inert environment for sensing elements, shielding them from degradation and environmental variations. This stability ensures the long-term functionality of biosensors, enabling consistent and reproducible measurements over extended durations. Additionally, aerogel-based biosensors can be regenerated through straightforward washing or regeneration procedures, allowing for multiple uses and reducing assay costs. This capability enhances the sustainability and cost-effectiveness of biosensing platforms, making them suitable for continuous monitoring applications. Furthermore, the ease of regeneration ensures minimal downtime between assays, enabling efficient workflow and increased throughput. Overall, the stability and regenerability of aerogel-based biosensors make them valuable tools for various analytical applications, including medical diagnostics, environmental monitoring, and food safety, where reliability, longevity, and cost efficiency are paramount (Naresh & Lee, 2021; Adetunji et al., 2023a).

Multiplexed Detection: Aerogels' versatility enables the integration of multiple sensing elements within a single platform, facilitating multiplexed detection of diverse analytes simultaneously. By functionalizing different regions of the aerogel scaffold with specific biomolecules, multiplexed biosensors can identify multiple targets in a single sample, enhancing throughput and efficiency in biomedical diagnostics. This approach allows for the simultaneous detection of various disease biomarkers or pathogens, providing comprehensive information from a single assay. Multiplexed biosensors based on aerogels offer advantages such as reduced sample volume, shorter assay times, and decreased costs compared to traditional single-analyte assays. Furthermore, the ability to detect multiple targets in a single test improves diagnostic accuracy and efficiency, making aerogel-based multiplexed biosensors invaluable tools for applications ranging from disease diagnosis and monitoring to drug discovery and personalized medicine. Overall, aerogel-enabled multiplexed biosensors hold great promise for advancing biomedical diagnostics and improving patient care (Glatz et al. 2022).

Point-of-Care Diagnostics: Aerogel-based biosensors hold significant promise for point-of-care diagnostics, offering rapid and decentralized testing capabilities suitable for resource-limited settings. Their portable and miniaturizable nature, combined with high sensitivity and specificity, positions them as ideal candidates for on-site detection of pathogens, biomarkers, and toxins. Aerogel-based biosensors enable timely diagnosis and treatment decisions by providing real-time results at the point of care, without the need for specialized laboratory equipment or trained personnel. This capability is particularly advantageous in remote or underserved areas where access to centralized diagnostic facilities is limited. Furthermore, aerogel-based biosensors offer potential applications in field epidemiology, outbreak surveillance, and disaster response scenarios, facilitating prompt and effective responses to public health threats. Overall, the portability, sensitivity, and ease of use of aerogel-based biosensors make them invaluable tools for improving healthcare delivery and disease management in diverse settings (Rasmi et al. 2021; Adetunji et al., 2023b).

Environmental Sensing: Beyond biomedical applications, aerogel-based biosensors are increasingly being employed in environmental monitoring to detect and quantify various pollutants, contributing to environmental sustainability and public health protection. By functionalizing aerogels with specific receptors, enzymes, or antibodies, they can selectively capture target analytes present in environmental samples. For instance, aerogels functionalized with heavy metal-binding ligands can selectively trap heavy metal ions such as lead, mercury, or cadmium. Similarly, aerogels modified with enzyme-based recognition elements can detect pesticides or organic pollutants by catalyzing specific chemical reactions. The high surface area and porous structure of aerogels enable efficient analyte capture and detection, while their stability and regenerability allow for repeated use in environmental monitoring applications. Aerogel-based biosensors offer advantages such as rapid response times, high sensitivity, and low detection limits, making them valuable tools for assessing environmental quality and identifying sources of contamination. By providing real-time data on pollutant levels, aerogel-based biosensors contribute to informed decision-making, pollution control measures, and the protection of ecosystems and human health. Overall, aerogel-based biosensors play a crucial role in advancing environmental monitoring efforts, promoting sustainability, and safeguarding the environment for future generations (Gavrila et al. 2022).

Furthermore, aerogels offer tremendous potential in biosensing applications, enabling sensitive, selective, and rapid detection of biological analytes for biomedical diagnostics, environmental monitoring, and other fields. With ongoing research and development efforts, aerogel-based biosensors are poised to revolutionize healthcare, environmental protection, and beyond, offering innovative solutions for a wide range of sensing challenges (Almeida et al. 2024).

8.4 IMAGING APPLICATIONS OF AEROGEL

Aerogels have promising applications in medical imaging, enhancing diagnostic accuracy and therapeutic efficacy (Table 2). Firstly, they serve as contrast agents in imaging techniques such as magnetic resonance imaging (MRI) and computed tomography (CT) scans, providing improved contrast for better visualization of tissues and organs. Secondly, aerogels can be utilized as drug delivery vehicles for imaging agents, ensuring targeted delivery to specific sites within the body and controlled release for enhanced imaging outcomes. Furthermore, aerogels with tailored optical properties find use in optical imaging techniques for biological samples, offering high transparency and light scattering properties. Lastly, aerogels can act as carriers for radiopharmaceuticals in nuclear imaging techniques like positron emission tomography (PET) scans, offering enhanced stability and controlled release of radiotracers. Overall, aerogels play a crucial role in advancing medical imaging technologies, enabling more accurate diagnoses and effective treatments (Hussain et al. 2022).

Table 2. Imaging applications of aerogel

Imaging Application	Description	Benefits	Examples	References
Contrast Agents	Aerogels can be used as contrast agents in medical imaging techniques such as MRI and CT scans.	Enhanced imaging contrast	Aerogel-based nanoparticles for MRI	Rahman, 2023.
Drug Delivery Vehicles	Aerogels can encapsulate imaging agents for targeted delivery to specific sites within the body.	Controlled release of imaging agents	Aerogel-based drug-loaded nanoparticles	Fijalkowski et al. 2023.
Optical Imaging	Aerogels with tailored optical properties can be used in optical imaging techniques for biological samples.	High transparency and light scattering properties	Aerogel-based optical imaging probes	Kim et al. 2019; Mathew et al., 2024b
Radiopharmaceutical Carriers	Aerogels can serve as carriers for radiopharmaceuticals used in nuclear imaging techniques like PET scans.	Enhanced stability and controlled release of radiotracers	Aerogel-based radiopharmaceutical carriers	Lau et al. 2020.

8.5 BONE REGENERATION APPLICATIONS OF AEROGELS

Bone regeneration presents a significant challenge in modern medicine, especially in cases of trauma, disease, or congenital defects. Aerogels, with their unique properties, offer promising solutions in this domain. In bone regeneration applications, aerogels serve as scaffolds that mimic the extracellular matrix, providing structural support and promoting cellular interactions crucial for tissue repair and regeneration (Zhang et al. 2023).

8.5.1 Properties of Aerogels in Bone Regeneration

High Porosity and Surface Area: Aerogels are characterized by their high porosity and surface area, making them ideal for tissue regeneration applications. This unique property enables efficient cellular infiltration, allowing cells to migrate and proliferate throughout the scaffold. Additionally, the large surface area of aerogels promotes enhanced nutrient exchange, facilitating the diffusion of essential molecules such as oxygen, growth factors, and nutrients to support cell metabolism and tissue growth. Furthermore, the porous structure of aerogels facilitates effective waste removal, ensuring the elimination of metabolic byproducts and maintaining a conducive microenvironment for tissue regeneration. Overall, the combination of high porosity and surface area in aerogels plays a crucial role in promoting cellular activities and supporting the physiological processes necessary for successful tissue regeneration (Carriero et al. 2023).

Tunable Mechanical Properties: The mechanical properties of aerogels are highly tunable, allowing for precise customization to match those of natural bone. This capability enables aerogels to provide adequate support during the healing process, enhancing their utility in orthopedic applications. By adjusting factors such as composition, density, and pore structure, aerogels can be engineered to exhibit desired mechanical characteristics, including stiffness, strength, and elasticity. This tailoring ensures that aerogel-based scaffolds offer sufficient structural integrity and load-bearing capacity to withstand physiological forces and promote proper tissue regeneration. Matching the mechanical properties of natural bone also minimizes the risk of stress shielding or implant failure, facilitating the integration of aerogel scaffolds into the surrounding tissue and improving overall clinical outcomes in bone repair and regeneration (Zhang et al.2023).

Biocompatibility: Aerogels demonstrate high biocompatibility, mitigating adverse reactions and fostering favorable interactions with surrounding tissues. This property is essential for their use in biomedical applications, ensuring compatibility with the biological environment and minimizing the risk of immune responses or tissue rejection. The inert nature of aerogels, combined with their porous structure and surface properties, facilitates cellular adhesion, proliferation, and integration into host tissues. As a result, aerogels offer a safe and effective platform for various biomedical applications, including tissue engineering, drug delivery, and regenerative medicine, where compatibility with biological systems is paramount for successful outcomes (Bakhori et al. 2023).

Osteoconductivity and Osteoinductivity: Aerogels can be precisely engineered to possess osteoconductive and osteoinductive properties, fostering the formation and differentiation of bone cells. Osteoconductive properties refer to the ability of a material to provide a scaffold for bone ingrowth, while osteoinductive properties involve the stimulation of precursor cells to differentiate into bone-forming cells. Through careful manipulation of composition, surface chemistry, and structural characteristics, aerogels can be tailored to mimic the natural extracellular matrix of bone tissue, promoting cell adhesion, proliferation, and differentiation. Additionally, the incorporation of bioactive molecules such as growth factors or osteogenic proteins further enhances the osteoinductive potential of aerogels. By combining osteoconductive and osteoinductive properties, aerogels create an environment conducive to bone regeneration, facilitating the formation of functional bone tissue. This capability makes aerogels valuable materials for various orthopedic applications, including bone defect repair, augmentation, and tissue engineering (Niu et al. 2023).

8.5.2 Applications in Bone Regeneration

Scaffold for Cell Delivery: Aerogels function as three-dimensional scaffolds for delivering bone-forming cells, such as mesenchymal stem cells (MSCs), to sites of bone defects. These scaffolds offer a supportive environment conducive to cell attachment, proliferation, and differentiation into osteoblasts, which are crucial for new bone formation. The porous structure of aerogels facilitates nutrient and oxygen diffusion to the encapsulated cells while providing mechanical support. Moreover, aerogels can be functionalized with bioactive molecules or extracellular matrix components to enhance cell adhesion, proliferation, and differentiation. This tailored environment promotes the formation of mineralized tissue and accelerates the bone healing process. By serving as carriers for bone-forming cells, aerogel scaffolds offer a promising approach for regenerating bone tissue in a controlled and localized manner. Their versatility and biocompatibility make them valuable tools in tissue engineering and regenerative medicine, with the potential to improve outcomes for patients with bone defects or injuries (Liu et al. 2023).

Drug Delivery Matrix: Aerogels serve as excellent carriers for growth factors, cytokines, and other bioactive molecules that facilitate bone regeneration. By loading these agents into the aerogel matrix, their controlled release is achieved, enhancing their efficacy and spatial localization at the defect site. This controlled release profile ensures a sustained presence of bioactive molecules, providing a continuous stimulus for cell proliferation, differentiation, and extracellular matrix synthesis, which are essential processes in bone healing. Furthermore, the ability to tailor the aerogel structure allows for precise control over the release kinetics, enabling the optimization of therapeutic outcomes. The localized delivery of growth factors and cytokines promotes tissue regeneration directly at the site of injury, minimizing systemic side effects and maximizing the therapeutic effect. Overall, the use of aerogels as delivery vehicles for bioactive molecules holds great promise in accelerating the bone healing process and improving clinical outcomes in bone regeneration therapies (Jiang et al. 2023).

Composite Materials: Aerogels can significantly enhance the mechanical properties of composite materials, making them suitable for load-bearing applications in bone regeneration. By integrating aerogels with materials like polymers, ceramics, or bioglasses, composite scaffolds with tailored properties can be engineered to meet specific clinical requirements. Aerogels, with their high porosity and low density, contribute to the lightweight nature of the composite while imparting excellent structural support. Moreover, the versatility of aerogels allows for precise tuning of mechanical properties such as stiffness, strength, and elasticity by adjusting factors like aerogel concentration, dispersion, and orientation within the composite matrix. This enables the fabrication of scaffolds that closely mimic the mechanical properties of natural bone, promoting optimal tissue regeneration and integration. Generally, the incorporation of aerogels into composite materials offers a promising approach to develop load-bearing scaffolds for bone regeneration, providing a customizable platform to address diverse clinical needs and challenges in orthopedic applications (Nallusamy & Das, 2021).

Guided Bone Regeneration: Aerogels offer significant potential in guided bone regeneration (GBR) techniques by serving as barrier membranes to promote selective tissue regeneration. In GBR, aerogels function as barriers to inhibit the infiltration of soft tissues into the defect site, creating a protected environment conducive to bone tissue regeneration. The high porosity and tunable properties of aerogels allow for the precise control of pore size, facilitating the passage of nutrients and oxygen while effectively blocking the ingress of soft tissue cells. By acting as barriers, aerogels prevent fibrous tissue ingrowth, which can interfere with proper bone regeneration and compromise the structural integrity of the newly formed bone. This selective exclusion of undesirable tissues allows bone cells to populate the defect site and initiate the process of bone formation without interference. Furthermore, aerogel-based barrier membranes can be tailored to degrade gradually over time, aligning with the rate of bone regeneration and providing continuous support until the defect is fully healed. In general, aerogels hold promise as effective barrier membranes in GBR techniques, offering precise control over tissue regeneration while minimizing complications and enhancing the success of bone regeneration procedures (Yang et al. 2022).

8.5.3 Challenges in Bone Regeneration with Aerogels

Biodegradation Rate: Controlling the biodegradation rate of aerogels is crucial in bone regeneration applications to ensure optimal tissue integration and structural support. If the aerogels degrade too rapidly, it may compromise the scaffold's structural integrity before sufficient new bone formation occurs, leading to implant failure. Conversely, inadequate degradation may impede tissue integration and hinder the remodeling process, resulting in poor functional outcomes. To address this challenge, researchers are exploring various approaches such as adjusting the aerogel composition, modifying cross-linking density, or incorporating biodegradable components to tailor the degradation kinetics. Additionally, designing aerogels with degradation profiles that align with the rate of new bone formation can promote successful tissue integration and regeneration. By carefully controlling the biodegradation rate, aerogel-based scaffolds can provide temporary support while facilitating the natural healing process, ultimately enhancing bone regeneration outcomes (Zhang et al. 2023).

Vascularization: Ensuring sufficient vascularization within regenerated tissue is crucial for its long-term viability and functionality. In the context of aerogel-based scaffolds for bone regeneration, promoting angiogenesis is essential to support nutrient delivery and waste removal processes vital for tissue survival. Several strategies can be employed to enhance angiogenesis within aerogel scaffolds. These include incorporating angiogenic growth factors such as vascular endothelial growth factor (VEGF) or fibroblast growth factor (FGF) into the scaffold matrix to stimulate blood vessel formation. Additionally, engineering the scaffold architecture to mimic the natural extracellular matrix and provide physical cues conducive to endothelial cell migration and vessel formation can promote angiogenesis. By facilitating the development of a functional vascular network within aerogel-based scaffolds, these strategies enhance nutrient perfusion and waste removal, thereby improving the long-term survival and integration of regenerated tissue (Simunovic & Finkenzeller, 2021).

Inflammatory Response: While aerogels are generally biocompatible, variations in composition and surface properties can trigger an inflammatory response in vivo, potentially hindering successful bone regeneration. To address this challenge, strategies must be developed to modulate the immune response and minimize inflammation. This may involve surface modification techniques to enhance biocompatibility and reduce immunogenicity, such as coating aerogels with bioactive molecules or polymers that promote tissue integration while mitigating inflammatory reactions. Additionally, incorporating anti-inflammatory agents or immunomodulatory factors into aerogel formulations can help regulate the immune response and create a more favorable microenvironment for bone regeneration. By mitigating inflammation and promoting tissue healing, these approaches enhance the overall biocompatibility and effectiveness of aerogel-based bone regeneration therapies (Batool et al. 2021).

Clinical Translation: Translating aerogel-based bone regeneration therapies from the laboratory to clinical practice necessitates navigating several key challenges. Firstly, overcoming regulatory hurdles is paramount, as aerogel products must comply with rigorous safety and efficacy standards set forth by regulatory authorities such as the FDA (Food and Drug Administration) or equivalent bodies in other regions. This involves extensive documentation, testing, and adherence to regulatory guidelines throughout the development and approval process. Secondly, standardizing production processes is essential to ensure consistency, quality, and scalability of aerogel-based therapies. Optimizing manufacturing methods, raw material sourcing, and quality control measures are crucial steps in achieving reproducible outcomes suitable for clinical use. Lastly, conducting rigorous preclinical and clinical evaluations is imperative to assess the safety, efficacy, and long-term outcomes of aerogel-based bone regeneration therapies in human subjects. Comprehensive preclinical studies in animal models followed by well-designed clinical trials are necessary to gather robust data on the therapeutic potential and performance of these treatments, ultimately paving the way for widespread clinical adoption and patient benefit (Bakhori et al. 2023).

Furthermore, aerogels hold great promise in bone regeneration applications due to their unique properties and versatile nature. By serving as scaffolds, drug delivery vehicles, and composite materials, aerogels facilitate tissue repair and regeneration in bone defects. However, addressing challenges such as biodegradation control, vascularization, inflammatory response modulation, and clinical translation is essential for realizing the full potential of aerogels in bone regeneration. With continued research and innovation, aerogel-based therapies have the potential to revolutionize the treatment of bone disorders and injuries, improving patient outcomes and quality of life (Lázár et al. 2023).

8.6 FUTURE PERSPECTIVES AND CHALLENGES

Aerogels have emerged as a promising class of materials with wide-ranging applications in biomedicine, including biosensing, imaging, and bone regeneration. Their unique properties, such as high surface area, tunable porosity, and biocompatibility, make them attractive candidates for various biomedical applications. However, as with any emerging technology, there are both future perspectives and challenges that need to be addressed to fully harness the potential of aerogels in healthcare (Nita et al. 2020).

8.6.1 Future Perspectives

Enhanced Functionalization: One of the key future perspectives in aerogel technology is the development of enhanced functionalization strategies. Tailoring aerogel surfaces for specific biomolecular interactions and functionalities could significantly broaden their applications in biosensing. By immobilizing biomolecules with precision, aerogels could become highly sensitive and selective biosensors for detecting a wide range of analytes, including biomarkers for diseases (Usala et al. 2023).

Multimodal Imaging Agents: Another exciting future perspective is the design of aerogels with integrated functionalities for multimodal imaging. Combining different imaging modalities, such as magnetic resonance imaging (MRI), computed tomography (CT), and fluorescence imaging, into a single aerogel platform could provide complementary information, enhancing diagnostic accuracy and facilitating personalized medicine (LuengoMorato et al. 2021).

Advanced Tissue Engineering: In the field of tissue engineering, aerogels hold great promise for advanced scaffold design. Future research could focus on expanding aerogel-based scaffolds to mimic the complex architectures of native tissues more accurately. By incorporating bioactive molecules, growth factors, and cells into aerogels, it may be possible to create tissue-engineered constructs that closely resemble natural tissues, promoting better regeneration and integration upon implantation (Ashammakhi et al. 2022).

Targeted Drug Delivery Systems: Aerogels offer unique opportunities for developing targeted drug delivery systems. By loading therapeutic agents into aerogel matrices and modifying their surface properties, it may be possible to achieve controlled and sustained release of drugs at specific target sites in the body. This approach could minimize systemic side effects and improve therapeutic outcomes for various diseases, including cancer and inflammatory disorders (García-González et al. 2021).

8.6.2 Challenges

Scalability of Production: One of the major challenges in the widespread adoption of aerogel-based biomedical technologies is the scalability of production. Current synthesis methods often involve complex and time-consuming processes that are not suitable for large-scale manufacturing. Developing scalable synthesis methods that can meet the growing demand for aerogel-based products is essential for their translation into clinical applications (Bakhori et al.2023).

Cost-Effectiveness: Cost is another significant barrier to the adoption of aerogel technology in healthcare. The production of aerogels typically involves expensive precursors and energy-intensive processes, resulting in high manufacturing costs. Addressing the cost barriers associated with aerogel production and functionalization is critical for making these technologies more accessible and affordable to patients and healthcare providers (Nguyen et al. 2022).

Biocompatibility: Ensuring the long-term biocompatibility of aerogels is essential for their safe use in biomedical applications. While aerogels are generally considered biocompatible, there is still a need for comprehensive biocompatibility studies to evaluate their interactions with different biological systems over extended periods. Minimizing adverse effects, such as inflammation and immune responses, is essential for the successful clinical translation of aerogel-based biomedical products (Neagu et al. 2023).

Regulatory Approval: Navigating regulatory pathways and obtaining approvals for aerogel-based biomedical products present significant challenges. Aerogels are considered novel materials, and their regulatory classification and requirements may vary depending on the intended application and jurisdiction. Overcoming regulatory hurdles and obtaining necessary approvals for aerogel-based medical devices, diagnostics, and therapeutics will require close collaboration between researchers, industry partners, and regulatory agencies (Boccia et al. 2023).

Generally, aerogels hold tremendous promise for revolutionizing biomedical applications such as biosensing, imaging, and bone regeneration. However, addressing the future perspectives and challenges outlined above will be crucial for unlocking their full potential in healthcare. With continued research, innovation, and collaboration, aerogels could soon become indispensable tools in diagnosing, treating, and managing various medical conditions, ultimately improving patient outcomes and quality of life (García-González et al. 2019).

8.7 CONCLUSION

In conclusion, aerogels represent a versatile class of materials with immense potential in various biomedical applications, including biosensing, imaging, and bone regeneration. Through their unique properties such as high surface area, tunable porosity, and biocompatibility, aerogels offer promising solutions to address critical challenges in healthcare. In biosensing, aerogels serve as efficient platforms for the immobilization of biomolecules, enabling sensitive detection of analytes through diverse detection mechanisms. Their application in imaging modalities as contrast agents and drug delivery vehicles holds great promise for enhancing diagnostic accuracy and therapeutic efficacy. Moreover, aerogels' ability to mimic the extracellular matrix and promote cellular interactions makes them ideal scaffolds for bone regeneration, facilitating tissue repair and regeneration. Despite significant progress, challenges such as scalability, cost-effectiveness, and long-term biocompatibility remain to be addressed. Future research efforts focused on optimizing aerogel synthesis techniques, tailoring their properties for specific biomedical applications, and conducting comprehensive preclinical and clinical studies will be crucial for realizing their full potential in healthcare. In addition, aerogels represent a promising class of materials that are poised to revolutionize biomedical technologies, offering innovative solutions for diagnostics, therapeutics, and tissue engineering, and paving the way for improved patient outcomes and quality of life.

REFERENCES

Adetunji, C. O., Egbuna, C., Ajiboye, A. A. A., Ajayi, O. O., Dauda, W. P., Ghazanfar, S., Ogundolie, F. A., Mathew, J. T., Adetunji, J. B., & Inobeme, A., OlotuTitilayo, Olaniyan, O. T., Ijabadeniyi, O. A., Aborode, A. T., Ficai, A., Akram, M., Bodunrinde, R. E., Laila, U., Olaniyan, O. P., Micheal, O. S., Aliyu, A., Akinbo, O. &Adetuyi, B. O. (2023)a. Biogenic nanomaterials with diverse biological activities in the food and biomedical industries, Evaporative Coolers for the Postharvest Management of Fruits and Vegetables, Academic Press, 395-420, . DOI: 10.1002/btm2.10124 PMID: 30680322

Adetunji, C. O., Mathew, J. T., Inobeme, A., Taiwo, O., Olaniyan, O., Shakira, G., & Wadazani, D.(2023)b.Application of Biosensors for Detection and Monitoring of Water Quality.In book: Organic and Inorganic Materials Based Sensors, John Wiley & Sons, Inc., 1065-1077, DOI: 10.1002/btm2.10124 PMID: 30680322

Afewerki, S., Sheikhi, A., Kannan, S., Ahadian, S., & Khademhosseini, A. (2018). Gelatin-polysaccharide composite scaffolds for 3D cell culture and tissue engineering: Towards natural therapeutics. *Bioengineering & Translational Medicine*, *4*(1), 96–115. DOI: 10.1002/btm2.10124 PMID: 30680322

Almeida, C. M. R., Merillas, B., & Pontinha, A. D. R. (2024). Trends on Aerogel-Based Biosensors for Medical Applications: An Overview. *International Journal of Molecular Sciences*, *25*(2), 1309. DOI: 10.3390/ijms25021309 PMID: 38279307

Ashammakhi, N., GhavamiNejad, A., Tutar, R., Fricker, A., Roy, I., Chatzistavrou, X., Hoque Apu, E., Nguyen, K.-L., Ahsan, T., Pountos, I., & Caterson, E. J. (2022). Highlights on Advancing Frontiers in Tissue Engineering. *Tissue Engineering. Part B, Reviews*, *28*(3), 633–664. DOI: 10.1089/ten.teb.2021.0012 PMID: 34210148

Bakhori, N. M., Ismail, Z., Hassan, M. Z., & Dolah, R. (2023). Emerging Trends in Nanotechnology: Aerogel-Based Materials for Biomedical Applications. *Nanomaterials (Basel, Switzerland), 13*(6), 1063. DOI: 10.3390/nano13061063 PMID: 36985957

Batool, F., Özçelik, H., Stutz, C., Gegout, P. Y., Benkirane-Jessel, N., Petit, C., & Huck, O. (2021). Modulation of immune-inflammatory responses through surface modifications of biomaterials to promote bone healing and regeneration. *Journal of Tissue Engineering, 12*, 20417314211041428. DOI: 10.1177/20417314211041428 PMID: 34721831

Boccia, A. C., Pulvirenti, A., García-González, C. A., Grisi, F., & Neagu, M. (2023). Compendium of Safety Regulatory for Safe Applications of Aerogels. *Gels (Basel, Switzerland), 9*(11), 842. DOI: 10.3390/gels9110842 PMID: 37998932

Carriero, V. C., Di Muzio, L., Petralito, S., Casadei, M. A., & Paolicelli, P. (2023). Cryogel Scaffolds for Tissue-Engineering: Advances and Challenges for Effective Bone and Cartilage Regeneration. *Gels (Basel, Switzerland), 9*(12), 979. DOI: 10.3390/gels9120979 PMID: 38131965

Fijalkowski, M., Ali, A., Qamer, S., Coufal, R., Adach, K., & Petrik, S. (2023). Hybrid and Single-Component Flexible Aerogels for Biomedical Applications: A Review. *Gels (Basel, Switzerland), 10*(1), 4. DOI: 10.3390/gels10010004 PMID: 38275842

García-González, C. A., Budtova, T., Durães, L., Erkey, C., Del Gaudio, P., Gurikov, P., Koebel, M., Liebner, F., Neagu, M., & Smirnova, I. (2019). An Opinion Paper on Aerogels for Biomedical and Environmental Applications. *Molecules (Basel, Switzerland), 24*(9), 1815. DOI: 10.3390/molecules24091815 PMID: 31083427

García-González, C. A., Sosnik, A., Kalmár, J., De Marco, I., Erkey, C., Concheiro, A., & Alvarez-Lorenzo, C. (2021). Aerogels in drug delivery: From design to application. Journal of controlled release: official journal of the Controlled Release Society, 332, 40–63. DOI: 10.3390/molecules24091815 PMID: 31083427

Gavrila , S., Ursachi, C. ., Per a-Cri an, S., & Munteanu, F. D. (2022). Recent Trends in Biosensors for Environmental Quality Monitoring. *Sensors (Basel), 22*(4), 1513. DOI: 10.3390/s22041513 PMID: 35214408

Glatz, R. T., Ates, H. C., Mohsenin, H., Weber, W., & Dincer, C. (2022). Designing electrochemical microfluidic multiplexed biosensors for on-site applications. *Analytical and Bioanalytical Chemistry, 414*(22), 6531–6540. DOI: 10.1007/s00216-022-04210-4 PMID: 35794347

Horvat, G., Pantić, M., Knez, Ž., & Novak, Z. (2022). A Brief Evaluation of Pore Structure Determination for Bioaerogels. *Gels (Basel, Switzerland), 8*(7), 438. DOI: 10.3390/gels8070438 PMID: 35877523

Hussain, S., Mubeen, I., Ullah, N., Shah, S. S. U. D., Khan, B. A., Zahoor, M., Ullah, R., Khan, F. A., & Sultan, M. A. (2022). Modern Diagnostic Imaging Technique Applications and Risk Factors in the Medical Field: A Review. *BioMed Research International, 5164970*(1), 5164970. Advance online publication. DOI: 10.1155/2022/5164970 PMID: 35707373

Inobeme, A., Adetunji, C. O., Mathew, J T., Ajai, A. I., Inobeme, J., Bamigboye, M. O., Onyeaku, S., Maliki, M., Eziukwu, C. &Tawa, K. (2023). Nanotechnology for Bioremediation of Heavy Metals. In book: Microbial Technologies in Industrial Wastewater Treatment. Springer Nature Singapore, 19-30. .DOI: 10.1007/978-981-99-2435-6_2

Inobeme, A., Mathew, J.T., Adetunji, C.O., Agbugui, M.O., Osarenren, E.J., Oti, C., Ngonso, B.F., Inobeme, J., Oyedolapo, M.B., Bernard, E. &Jibrin, N.A. (2024). Biocomposites with graphene derivatives. In book: Advances in Biocomposites and their Applications, Elsevier Inc., 149-166. DOI: DOI: 10.1016/B978-0-443-19074-2.00005-8

Jiang, Z., Zheng, Z., Yu, S., Gao, Y., Ma, J., Huang, L., & Yang, L. (2023). Nanofiber Scaffolds as Drug Delivery Systems Promoting Wound Healing. *Pharmaceutics, 15*(7), 1829. DOI: 10.3390/pharmaceutics15071829 PMID: 37514015

Khan, N. R., Sharmin, T., & Bin Rashid, A. (2023). Exploring the Versatility of Aerogels: Broad Applications in Biomedical Engineering, Astronautics, Energy Storage, Biosensing, and Current Progress. *Heliyon, 10*(1), e23102. DOI: 10.1016/j.heliyon.2023.e23102 PMID: 38163169

Kim, Y., Baek, S., Gupta, P., Kim, C., Chang, K., Ryu, S. P., Kang, H., Kim, W. S., Myoung, J., Park, W., & Kim, K. (2019). Air-like plasmonics with ultralow-refractive-index silica aerogels. *Scientific Reports*, *9*(1), 2265. DOI: 10.1038/s41598-019-38859-2 PMID: 30783170

Lau, J., Rousseau, E., Kwon, D., Lin, K. S., Bénard, F., & Chen, X. (2020). Insight into the Development of PET Radiopharmaceuticals for Oncology. *Cancers (Basel)*, *12*(5), 1312. DOI: 10.3390/cancers12051312 PMID: 32455729

Lázár, I., Čelko, L., & Menelaou, M. (2023). Aerogel-Based Materials in Bone and Cartilage Tissue Engineering-A Review with Future Implications. *Gels (Basel, Switzerland)*, *9*(9), 746. DOI: 10.3390/gels9090746 PMID: 37754427

Le, C. V., & Yoon, H. (2024). Advances in the Use of Conducting Polymers for Healthcare Monitoring. *International Journal of Molecular Sciences*, *25*(3), 1564. DOI: 10.3390/ijms25031564 PMID: 38338846

Liu, J., Yang, L., Liu, K., & Gao, F. (2023). Hydrogel scaffolds in bone regeneration: Their promising roles in angiogenesis. *Frontiers in Pharmacology*, *14*, 1050954. DOI: 10.3389/fphar.2023.1050954 PMID: 36860296

Luengo Morato, Y., Ovejero Paredes, K., Lozano Chamizo, L., Marciello, M., & Filice, M.LuengoMorato. (2021). Recent Advances in Multimodal Molecular Imaging of Cancer Mediated by Hybrid Magnetic Nanoparticles. *Polymers*, *13*(17), 2989. DOI: 10.3390/polym13172989 PMID: 34503029

Mathew, J. T., Inobeme, A., Abdulkadir, A., Etsuyankpa, M. B., Shaba, E. Y., Musah, M., Azeh, Y., Adetunji, C. O., Ibrahim, A. M., Tanko, E., & Mamman, A. (2024). a. Properties and Characteristics of Nanocoatings for Medicinal Applications. In book: Sustainable Approach to Protective Nanocoatings. IGI Global; publishing Tomorrow's. *Advances in Chemical and Materials Engineering*, •••, 203–227. DOI: 10.4018/979-8-3693-3136-1.ch007

Mathew, J. T., Inobeme, A., Musah, M., Azeh, Y., Abdullahi, A., Shaba, E. Y., Salihu, A. M., Muhammad, E. B., Josiah, J. G., Jibrin, N. A., Ismail, H., Muhammad, A. I., Maurice, J., Mamman, A., & Ndamitso, M. M. (2024). b. A Critical Review of Green Approach on Wastewater Treatment Strategies. *Journal of Applied Science & Environmental Management*, *28*(2), 363–391. DOI: 10.4314/jasem.v28i2.9

Mathew, J. T., Musah, M., Azeh, Y., & Muhammed, M. (2023). a. Kinetic Study of Heavy Metals Removal from Pharmaceutical Wastewater Using Geopolymer/Fe_3O_4Nanocomposite. *Bima Journal of Science and Technology*, *7*(4), 152–163. DOI: 10.56892/bima.v7i4.539

Mathew, J. T., Musah, M., Azeh, Y., & Muhammed, M. (2023). b. Adsorptive Removal of Selected Toxic Metals from Pharmaceutical Wastewater using Fe_3O_4/ZnONanocomposite. *Dutse Journal of Pure and Applied Sciences*, *9*(4a), 236–248. DOI: 10.4314/dujopas.v9i4a.22

Nallusamy, J., & Das, R. K. (2021). Hydrogels and Their Role in Bone Tissue Engineering: An Overview. *Journal of Pharmacy & Bioallied Sciences*, *13*(Suppl 2), S908–S912. DOI: 10.4103/jpbs.jpbs_237_21 PMID: 35017896

Naresh, V., & Lee, N. (2021). A Review on Biosensors and Recent Development of Nanostructured Materials-Enabled Biosensors. *Sensors (Basel)*, *21*(4), 1109. DOI: 10.3390/s21041109 PMID: 33562639

Neagu, M., Grisi, F., Pulvirenti, A., Simón-Vázquez, R., García-González, C. A., & Boccia, A. C. (2023). Updated Aspects of Safety Regulations for Biomedical Applications of Aerogel Compounds—Compendia-Like Evaluation. *Safety (Basel, Switzerland)*, *9*(4), 80. DOI: 10.3390/safety9040080

Nguyen, P. T. T., Do, N. H. N., Goh, X. Y., Goh, C. J., Ong, R. H., Le, P. K., Phan-Thien, N., & Duong, H. M. (2022). Recent Progresses in Eco-Friendly Fabrication and Applications of Sustainable Aerogels from Various Waste Materials. *Waste and Biomass Valorization*, *13*(4), 1825–1847. DOI: 10.1007/s12649-021-01627-3 PMID: 34745389

Nita, L. E., Ghilan, A., Rusu, A. G., Neamtu, I., & Chiriac, A. P. (2020). New Trends in Bio-Based Aerogels. *Pharmaceutics*, *12*(5), 449. DOI: 10.3390/pharmaceutics12050449 PMID: 32414217

Niu, Y., Chen, L., & Wu, T. (2023). Recent Advances in Bioengineering Bone Revascularization Based on Composite Materials Comprising Hydroxyapatite. *International Journal of Molecular Sciences*, *24*(15), 12492. DOI: 10.3390/ijms241512492 PMID: 37569875

Noman, M. T., Amor, N., Ali, A., Petrik, S., Coufal, R., Adach, K., & Fijalkowski, M. (2021). Aerogels for Biomedical, Energy and Sensing Applications. *Gels (Basel, Switzerland)*, *7*(4), 264. DOI: 10.3390/gels7040264 PMID: 34940324

Rahman, M. (2023). Magnetic Resonance Imaging and Iron-oxide Nanoparticles in the era of Personalized Medicine. *Nanotheranostics*, *7*(4), 424–449. DOI: 10.7150/ntno.86467 PMID: 37650011

Rasmi, Y., Li, X., Khan, J., Ozer, T., & Choi, J. R. (2021). Emerging point-of-care biosensors for rapid diagnosis of COVID-19: Current progress, challenges, and future prospects. *Analytical and Bioanalytical Chemistry*, *413*(16), 4137–4159. DOI: 10.1007/s00216-021-03377-6 PMID: 34008124

rtunities. Advanced science (Weinheim, Baden-Wurttemberg, Germany), 10(9), e2205762. DOI: 10.1002/advs.202205762

Sheng, Z., Liu, Z., Hou, Y., Jiang, H., Li, Y., Li, G., & Zhang, X. (2023). *The Rising Aerogel Fibers: Status*. Challenges, and Oppo.

Simunovic, F., & Finkenzeller, G. (2021). Vascularization Strategies in Bone Tissue Engineering. *Cells*, *10*(7), 1749. DOI: 10.3390/cells10071749 PMID: 34359919

Smith, S., Goodge, K., Delaney, M., Struzyk, A., Tansey, N., & Frey, M. (2020). A Comprehensive Review of the Covalent Immobilization of Biomolecules onto ElectrospunNanofibers. *Nanomaterials (Basel, Switzerland)*, *10*(11), 2142. DOI: 10.3390/nano10112142 PMID: 33121181

Thie, C., Quallen, S., Ibrahim, A., Xing, T., & Johnson, B. (2023). Study of Energy Saving Using Silica Aerogel Insulation in a Residential Building. *Gels (Basel, Switzerland)*, *9*(2), 86. DOI: 10.3390/gels9020086 PMID: 36826255

Usala, E., Espinosa, E., El Arfaoui, W., Morcillo-Martín, R., Ferrari, B., & González, Z. (2023). Antibacterial Aerogels-Based Membranes by Customized Colloidal Functionalization of TEMPO-Oxidized Cellulose Nanofibers Incorporating CuO. *Bioengineering (Basel, Switzerland)*, *10*(11), 1312. DOI: 10.3390/bioengineering10111312 PMID: 38002436

Yang, Z., Wu, C., Shi, H., Luo, X., Sun, H., Wang, Q., & Zhang, D. (2022). Advances in Barrier Membranes for Guided Bone Regeneration Techniques. *Frontiers in Bioengineering and Biotechnology*, *10*, 921576. DOI: 10.3389/fbioe.2022.921576 PMID: 35814003

Zhang, L., Lei, Y., He, P., Wu, H., Guo, L., & Wei, G. (2022). Carbon Material-Based Aerogels for Gas Adsorption: Fabrication, Structure Design, Functional Tailoring, and Applications. *Nanomaterials (Basel, Switzerland)*, *12*(18), 3172. DOI: 10.3390/nano12183172 PMID: 36144967

Zhang, Y., Jiang, S., Xu, D., Li, Z., Guo, J., Li, Z., & Cheng, G. (2023). Application of Nanocellulose-Based Aerogels in Bone Tissue Engineering: Current Trends and Outlooks. *Polymers*, *15*(10), 2323. DOI: 10.3390/polym15102323 PMID: 37242898

Zheng, L., Zhang, S., Ying, Z., Liu, J., Zhou, Y., & Chen, F. (2020). Engineering of Aerogel-Based Biomaterials for Biomedical Applications. *International Journal of Nanomedicine*, *15*, 2363–2378. DOI: 10.2147/IJN.S238005 PMID: 32308388

Chapter 6
Catalytic Application of Aerogels

John Tsado Mathew
https://orcid.org/0000-0001-9514-5330
Department of Chemistry, Ibrahimm Badamasi Babangida Lapai, Nigeria

Musah Monday
https://orcid.org/0000-0002-3224-1664
Department of Chemistry, Ibrahim Badamasi Babangida University, Lapai, Nigeria

Yakubu Azeh
https://orcid.org/0000-0002-3818-674X
Department of Chemistry, Ibrahim Badamasi Babangida University, Lapai, Nigeria

Jonathan Hussaini
https://orcid.org/0009-0004-9216-8691
Department of Chemistry, Ibrahim Badamasi Babangida University, Lapai, Nigeria

Charles Oluwaseun Adetunji
Applied Microbiology, Biotechnology, and Nanotechnology Laboratory, Department of Microbiology, Edo State University, Nigeria

Musa Safiyanu Tanko
Department of Chemistry, Federal University Lafia, Nigeria

ABSTRACT

Aerogels have emerged as promising catalyst supports due to their high surface area, tunable porosity, and tailored surface chemistry. This review explores the catalytic application of aerogels in various reactions, including heterogeneous catalysis, photocatalysis, and enzymatic catalysis.

DOI: 10.4018/979-8-3693-5971-6.ch006

Aerogel-based catalysts exhibit enhanced catalytic activity, selectivity, and stability compared to traditional catalysts. The unique properties of aerogels, such as their high surface-to-volume ratio and uniform pore structure, facilitate efficient mass transfer and active site accessibility, leading to improved reaction kinetics. Moreover, the ability to functionalize aerogel surfaces enables the immobilization of catalytic species, further enhancing catalytic performance. This review highlights recent advances and future prospects in the field of aerogel-based catalysis, emphasizing the significance of aerogels as versatile and effective catalyst supports for a wide range of industrial and environmental applications

4.1 INTRODUCTION

Aerogels, with their remarkable properties such as high surface area, tunable pore structure, thermal stability, and chemical inertness, have emerged as promising materials for a wide range of catalytic applications. These unique characteristics make aerogels highly attractive as support materials for catalysts, enabling efficient catalytic reactions in diverse chemical processes. In this review, we delve into the catalytic applications of aerogels, highlighting their key advantages and recent developments in various fields (Bakhori et al. 2023).

Aerogel-based catalysts play a crucial role in environmental remediation, offering efficient solutions for the purification of air and water. In air pollution control, aerogel-supported catalysts are utilized in catalytic converters to reduce emissions of harmful gases from vehicles and industrial sources. Transition metal catalysts supported on aerogels effectively catalyze the conversion of nitrogen oxides (NOx), carbon monoxide (CO), and hydrocarbons into less harmful compounds such as nitrogen, carbon dioxide, and water vapor. In water treatment applications, aerogel-based catalysts are functionalized to remove pollutants and contaminants from wastewater sources. Metal or metal oxide nanoparticles immobilized on the aerogel surface catalyze the degradation of organic pollutants through advanced oxidation processes, ensuring efficient removal of harmful substances before discharge into the environment (Khan et al. 2023).

Aerogel-based catalysts play a crucial role in energy conversion and storage technologies, contributing to the development of sustainable energy systems. In hydrogen production, aerogels serve as catalyst supports in various processes such as steam reforming, water-gas shift reaction, and photocatalytic water splitting. Transition metal catalysts dispersed on aerogels facilitate the efficient conversion of fossil fuels or renewable feedstocks into hydrogen gas, a clean and versatile energy carrier. Moreover, aerogel-supported catalysts are employed in fuel cells for electrochemical energy conversion. Platinum-based catalysts supported on aerogels enhance the kinetics of oxygen reduction reactions at the cathode, improving the performance and durability of fuel cell systems. Additionally, aerogel-based catalysts are explored for electrocatalytic processes such as oxygen evolution reaction (OER) and hydrogen evolution reaction (HER), crucial for the development of electrolyzers and energy storage devices (Liu et al. 2023).

Aerogel-based catalysts find applications in petrochemical refining processes for the production of fuels and chemicals. Supported metal catalysts on aerogels enable selective hydrogenation, dehydrogenation, and hydrocarbon cracking reactions, essential for upgrading crude oil fractions and synthesizing value-added products. The high surface area and thermal stability of aerogels enhance catalytic performance under harsh operating conditions encountered in petrochemical reactors. Moreover, aerogel-supported catalysts are employed in Fischer-Tropsch synthesis for the production of liquid hydrocarbons from syngas (mixture of carbon monoxide and hydrogen). Transition metal catalysts dispersed on aerogels promote the formation of long-chain hydrocarbons, contributing to the synthesis of fuels and specialty chemicals from renewable or fossil-derived feedstocks (Khandelwal et al. 2023).

Aerogel-based catalysts are utilized in fine chemical synthesis for the production of pharmaceuticals, agrochemicals, and specialty materials. Supported metal catalysts on aerogels facilitate various organic transformations such as hydrogenation, oxidation, and C-C bond formation, enabling the efficient synthesis of complex molecules with high yield and selectivity. Furthermore, aerogel-supported catalysts are explored in asymmetric catalysis for the production of chiral compounds with pharmaceutical importance. Chiral ligands immobilized on aerogels serve as catalysts for asymmetric hydrogenation or asymmetric transfer hydrogenation reactions, enabling the synthesis of enantiopure compounds for pharmaceutical applications (Lermontov et al. 2023).

In addition, aerogels have emerged as versatile and effective support materials for catalysis, offering numerous advantages in environmental, energy, petrochemical, and fine chemical applications. The unique properties of aerogels, including high surface area, thermal stability, chemical inertness, and tunable surface chemistry, make them highly attractive for a wide range of catalytic reactions. Continuous research efforts are focused on further enhancing the catalytic performance and expanding the applications of aerogel-based catalysts to address emerging challenges in sustainable development and green chemistry (Zhang et al. 2022). The primary focus of this chapter is on aerogel based catalysis (ABC). It highlights the concept of catalysis, types and merits of aerogel based catalysis and their applications (figure 1).

Figure 1. Schematic representation of chapter

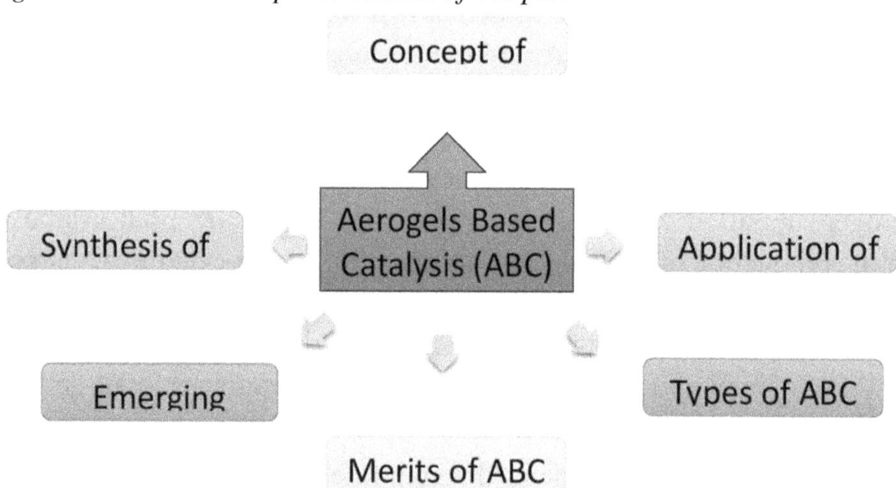

4.2 COMPARISON OF AEROGELS WITH OTHER SUPPORT MATERIALS

Aerogels, as support materials for catalysts, possess unique characteristics that set them apart from traditional support materials such as zeolites, mesoporous materials, and porous carbons. In this comparison, we'll explore the main differences and similarities between aerogels and these support materials (Budtova et al. 2020).

Surface Area and Pore Structure: Aerogels are renowned for their exceptionally high surface area and nanoscale pore structure, providing a large number of active sites for catalytic reactions. This high surface area and pore volume enhance mass transport and facilitate efficient utilization of active species. Zeolites exhibit high surface area and well-defined microporous structures with uniform pore sizes. While they offer excellent shape selectivity and molecular sieving properties, their pore sizes may limit diffusion of larger reactant molecules. Mesoporous materials possess intermediate pore sizes (2-50 nm) between microporous zeolites and macroporous materials. They offer improved mass transport compared to zeolites but may not match the surface area of aerogels. Porous carbons have a diverse range of pore sizes and structures depending on their synthesis method. While they can achieve high surface areas, their pore structures may not be as well-defined as aerogels, limiting their application in certain catalytic reactions (Franco et al. 2021).

Thermal Stability: Aerogels exhibit high thermal stability, making them suitable for catalytic applications under harsh operating conditions, including elevated temperatures. Their amorphous silica or carbon-based frameworks withstand high temperatures without structural degradation. Zeolites are thermally stable materials, but some zeolite frameworks may undergo structural changes or collapse at high temperatures, affecting their catalytic performance. Depending on their composition and synthesis method, mesoporous materials can exhibit varying degrees of thermal stability. Some mesoporous silica materials may exhibit good thermal stability similar to aerogels. Porous carbons generally exhibit good thermal stability, although their stability can vary depending on the precursor material and synthesis conditions (Jin et al. 2023).

Chemical Inertness: Aerogels are chemically inert materials, resistant to corrosion, oxidation, and chemical leaching. This inert behavior ensures stability and longevity of catalytic performance in diverse reaction conditions. Zeolites are also chemically stable materials, but certain acidic or alkaline conditions may affect their stability and catalytic activity over time. Like aerogels, some mesoporous materials exhibit chemical inertness, although their stability may vary depending on the specific composition and surface functionalization. Porous carbons generally possess good chemical stability, but they may be susceptible to oxidation or chemical attack under certain conditions (Wang et al. 2020; Mathew et al., 2024a).

Functionalization and Modification: Aerogels offer opportunities for surface functionalization and modification to introduce specific catalytic functionalities or enhance catalytic properties. Chemical functionalization, doping, or deposition of metal nanoparticles onto the aerogel surface can improve catalytic activity, selectivity, and durability. Zeolites can also be functionalized or modified through ion exchange, impregnation, or covalent attachment of catalytic species. However, their pore structure may limit the accessibility of functional groups or active sites. Mesoporousmaterials can be functionalized with various functional groups or metal nanoparticles to tailor their catalytic properties. Their larger pore sizes compared to zeolites may offer advantages in functionalization. Porous carbons are highly versatile materials for functionalization and modification due to their diverse surface chemistry. Functional groups can be introduced via chemical treatments or synthesis methods, allowing for precise control over catalytic properties (Maio et al. 2021; Inobeme et al., 2023).

Cost and Scalability: Aerogel production techniques have advanced significantly, enabling scalable and cost-effective manufacturing processes. However, aerogels may still be more expensive than some traditional support materials. Zeolites are widely available commercially and are relatively cost-effective compared to aerogels. Their production processes are well-established and can be scaled up for large-scale industrial applications. Mesoporous materials can be synthesized using relatively simple and cost-effective methods, making them attractive for large-scale production. The cost of porous carbons depends on the precursor material and synthesis method. Some carbon precursors may be inexpensive, but additional processing steps may increase production costs (Shan et al. 2024).

Furthermore, aerogels offer unique advantages as support materials for catalysts, including high surface area, thermal stability, chemical inertness, and tunable surface chemistry. While traditional support materials such as zeolites, mesoporous materials, and porous carbons also possess desirable properties, aerogels stand out for their exceptional surface characteristics and versatility in catalytic applications (Sun et al. 2023).

4.3 TYPES OF CATALYTIC REACTIONS

Various catalytic reactions include hydrogenation (addition of hydrogen), oxidation (removal of electrons), dehydrogenation (removal of hydrogen), isomerization (rearrangement of atoms), and hydrolysis (bond cleavage with water) (**Table 1**). These reactions are crucial in diverse industries such as petrochemicals, pharmaceuticals, and food processing. Catalysis accelerates these reactions by providing an alternative reaction pathway with lower activation energy, often facilitated by transition metal catalysts, enzymes, or zeolites (Hayes et al. 2023).

Table 1. Types of catalytic reactions

Type of Catalytic Reaction	Description	Example	Catalyst Used	References
Hydrogenation	Addition of hydrogen to unsaturated compounds to form saturated compounds.	Hydrogenation of alkenes to alkanes	Transition metal catalysts	Gao et al. 2021.
Oxidation	Removal of electrons from a substance, often resulting in the addition of oxygen.	Oxidation of alcohols to aldehydes or ketones	Metal oxides, enzymes	Guengerich and Yoshimoto, 2018; Inobeme et al., 2024
Dehydrogenation	Removal of hydrogen from a substance.	Dehydrogenation of alkanes to alkenes	Transition metal catalysts	McKay et al. 2019; Adetunji et al., 2023

continued on following page

Table 1. Continued

Type of Catalytic Reaction	Description	Example	Catalyst Used	References
Isomerization	Rearrangement of atoms within a molecule to form isomers.	Isomerization of n-butane to isobutane	Zeolite catalysts	Potter et al. 2020.
Hydrolysis	Chemical reaction involving the cleavage of a bond with water.	Hydrolysis of esters to carboxylic acids	Acid or base catalysts	Tebes-Stevens et al. 2017; Mathew et al., 2023a

4.4 ADVANTAGES OF AEROGEL-BASED CATALYSTS

Aerogel-based catalysts offer a myriad of advantages (figure 2) that make them attractive for a wide range of applications in catalysis, environmental remediation, energy production, and more. Here, we'll delve into the key advantages of aerogel-based catalysts (Obeid & Younes, 2024).

Figure 2. Advantages of aerogel-based catalysis

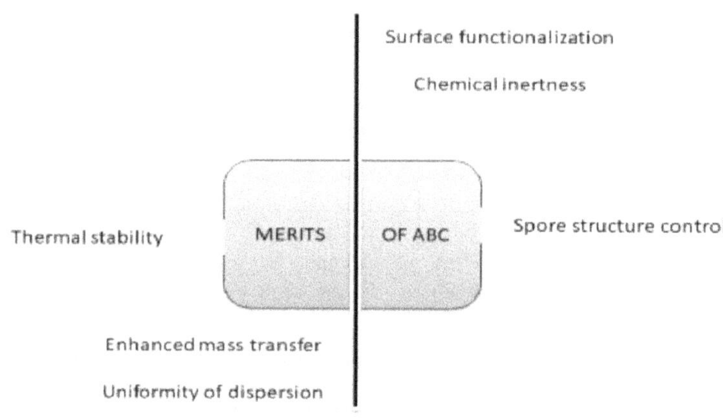

High Surface Area: One of the most significant advantages of aerogel-based catalysts lies in their remarkable surface area. Aerogels boast nanoscale pores and interconnected structures, which contribute to an exceptionally large surface area per unit volume. This unique characteristic facilitates a higher density of active sites available for catalytic reactions. With more active sites accessible within the material, aerogel-based catalysts exhibit enhanced catalytic activity compared to traditional catalysts. This increased activity enables more efficient conversion of reactants into desired products, thereby improving overall process efficiency and productivity. The large surface area of aerogel-based catalysts also promotes better interaction between reactant molecules and catalytic sites, leading to enhanced selectivity and specificity in catalytic transformations. As a result, aerogel-based catalysts hold great promise for a wide range of catalytic applications across various industries, including environmental remediation, energy production, and chemical synthesis (Bakhori et al. 2023).

Pore Structure Control: Aerogel synthesis methods provide researchers with a powerful toolkit for precisely controlling the pore structure and morphology of the material. Through careful adjustment of synthesis parameters such as precursor concentrations, solvent composition, and drying conditions, scientists can tailor the pore size, pore volume, and pore distribution of aerogels to meet specific catalytic requirements. This tunability is crucial for optimizing the catalytic performance and selectivity of aerogel-based catalysts. By modulating precursor concentrations, researchers can influence the density of the gel network formed during the sol-gel process, thereby controlling the overall pore structure of the aerogel. Additionally, varying the solvent composition and drying conditions allows for the manipulation of solvent evaporation rates and gelation kinetics, leading to precise control over pore size and morphology (Sozcu et al. 2023).

The ability to fine-tune the pore structure of aerogel-based catalysts enables optimization of key catalytic parameters such as mass transport, surface area, and accessibility of active sites. This optimization enhances catalytic efficiency and selectivity, leading to improved performance in various catalytic applications. Overall, the tunability of aerogel synthesis methods offers researchers unprecedented control over the structural properties of aerogel-based catalysts, enabling tailored design for specific catalytic processes and applications (Bari et al. 2024).

Uniform Dispersion of Active Sites: Aerogel-based catalysts excel in facilitating the uniform dispersion of active catalytic species throughout the material matrix. This uniform distribution ensures the efficient utilization of active sites, minimizing mass transfer limitations commonly encountered in heterogeneous catalysis. By evenly dispersing catalytic species within the aerogel matrix, the accessibility of reactants to active sites is enhanced, promoting more effective catalytic interactions and reaction kinetics. Furthermore, the uniform dispersion of active species helps to mitigate localized concentration gradients, preventing the occurrence of hotspots and ensuring consistent catalytic performance across the entire catalyst volume. As a result, aerogel-based catalysts exhibit improved catalytic efficiency and stability, making them highly attractive for a wide range of industrial applications, including environmental remediation, energy production, and chemical synthesis (Wang et al. 2023).

Enhanced Mass Transport: The interconnected pore structure of aerogels plays a pivotal role in promoting efficient mass transport of reactants and products within the catalyst. This unique characteristic allows for the rapid diffusion of reactant molecules to active sites and facilitates the efficient removal of product molecules, ultimately enhancing catalytic efficiency and reaction kinetics. The interconnected network of pores in aerogels provides continuous pathways for the transportation of molecules throughout the catalyst material. As reactant molecules penetrate the porous structure, they encounter a high density of active sites distributed uniformly within the aerogel matrix. This proximity to active sites enables fast and effective catalytic transformations, leading to enhanced reaction rates (Luo et al. 2023).

Additionally, the interconnected pores facilitate the swift removal of product molecules from the reaction environment, preventing their accumulation and minimizing undesirable side reactions. This efficient removal of products helps maintain favorable reaction conditions and promotes sustained catalytic activity over prolonged reaction times. Overall, the interconnected pore structure of aerogel-based catalysts enables rapid and unhindered mass transport, allowing for efficient diffusion of reactants to active sites and effective removal of products. This enhanced mass transport contributes to improved catalytic efficiency and reaction kinetics, making aerogel-based catalysts valuable tools for a wide range of chemical processes and industrial applications (Nghiem et al. 2020; Mathew et al., 2024b).

High Thermal Stability: Aerogels possess remarkable thermal stability, rendering them well-suited for catalytic applications in demanding operating conditions. The amorphous silica or carbon-based frameworks of aerogels can withstand high temperatures without undergoing structural degradation or compromising catalytic activity. This inherent thermal stability is attributed to the covalent bonds within the aerogel matrix, which remain intact even at elevated temperatures. As a result, aerogel-based catalysts maintain their structural integrity and catalytic performance under harsh operating conditions, including high temperatures encountered in various catalytic processes. This thermal stability enables aerogel-based catalysts to exhibit prolonged durability and reliability, ensuring consistent performance over extended reaction periods. Furthermore, the ability of aerogel-based catalysts to retain their catalytic activity at high temperatures opens up opportunities for catalytic applications that require elevated operating temperatures, such as steam reforming, combustion reactions, and thermal decomposition processes. Overall, the high thermal stability of aerogels enhances the versatility and applicability of aerogel-based catalysts across a broad spectrum of catalytic applications in diverse industrial sectors (Arshad et al. 2023).

Chemical Inertness: Aerogels' chemical inertness renders them compatible with a broad array of chemical environments. Their resistance to corrosion, oxidation, and chemical leaching ensures the stability and longevity of catalytic performance across diverse reaction conditions. This inert behavior arises from the stable chemical bonds present in the aerogel matrix, which remain unaffected by chemical interactions. In catalytic applications, where exposure to various reactants and products is common, aerogel-based catalysts offer unparalleled durability and reliability. Their resistance to chemical attack minimizes the risk of catalyst degradation, allowing for sustained performance over prolonged operating periods. Additionally, the chemical inertness of aerogels makes them suitable for use in aggressive reaction environments, including acidic or basic solutions, high-temperature conditions, and corrosive atmospheres. This versatility makes aerogel-based catalysts valuable tools for a wide range of catalytic processes in industries such as petrochemicals, environmental remediation, and renewable energy (Anžlovar & Žagar, 2022).

Tailorable Surface Chemistry: The surface chemistry of aerogels is highly tunable, allowing for precise modifications to introduce specific catalytic functionalities or enhance catalytic properties. Surface modification techniques such as chemical functionalization, doping, or deposition of metal nanoparticles enable the incorporation of catalytically active species onto the aerogel surface, leading to improved catalytic activity, selectivity, and durability. Chemical functionalization involves the covalent attachment of functional groups onto the aerogel surface, allowing for the introduction of specific catalytic sites or enhancement of surface reactivity. Doping involves the incorporation of dopant atoms or ions into the aerogel matrix, altering its electronic structure and catalytic behavior. Deposition of metal nanoparticles onto the aerogel surface provides catalytically active sites for various chemical reactions, enhancing catalytic activity and selectivity. These surface modification techniques offer precise control over the composition, structure, and surface properties of aerogel-based catalysts, enabling tailored design for specific catalytic applications. By optimizing the surface chemistry of aerogels, researchers can enhance their catalytic performance, expand their applicability, and address specific challenges in catalysis, environmental remediation, and energy conversion processes. Overall, the tunable surface chemistry of aerogels makes them versatile platforms for advanced catalytic materials with tailored functionalities and enhanced performance (Abbas et al. 2022).

Versatility and Multifunctionality: Aerogel-based catalysts serve as versatile platforms adaptable for diverse catalytic applications and reaction types. They find utility in heterogeneous catalysis, photocatalysis, electrocatalysis, and enzymatic catalysis, among others. Moreover, aerogel-based catalysts can exhibit multifunctional properties, integrating catalytic activity with functionalities like adsorption, sensing, or separation. This versatility expands their utility in complex chemical processes and environmental applications. In heterogeneous catalysis, aerogel-based catalysts facilitate chemical reactions at the interface between solid catalysts and liquid or gas phases. In photocatalysis, they harness light energy to drive catalytic reactions for environmental remediation or solar energy conversion. In electrocatalysis, aerogel-based materials facilitate electrochemical reactions, such as fuel cell reactions or water electrolysis, for energy conversion and storage. Additionally, in enzymatic catalysis, aerogel matrices provide a supportive environment for immobilized enzymes, enabling selective and efficient biochemical transformations. The multifunctional nature of aerogel-based catalysts makes them valuable tools for addressing diverse challenges in chemical synthesis, pollution control, and sustainable energy production (Khan et al. 2023).

Green and Sustainable: Aerogel synthesis is characterized by the use of environmentally benign precursors and solvents, rendering aerogels inherently green and sustainable materials. The sol-gel process employed in aerogel synthesis consumes less energy compared to conventional manufacturing methods, contributing to their eco-friendliness. Additionally, the use of renewable or recyclable raw materials further enhances the sustainability of aerogel production. By utilizing non-toxic precursors and solvents, aerogel synthesis minimizes environmental impact and reduces potential health risks associated with chemical manufacturing processes. Moreover, the sol-gel method allows for the production of aerogels at relatively low temperatures, conserving energy and reducing carbon emissions. Furthermore, incorporating renewable or recyclable raw materials into aerogel synthesis promotes circular economy principles and reduces reliance on finite resources. Overall, the eco-friendly attributes of aerogel-based materials make them attractive options for sustainable applications in various industries, including construction, energy storage, and environmental remediation (Noman et al. 2021; Mathew et al., 2023b).

Scalability and Cost-Effectiveness: Significant advancements in aerogel production techniques have led to scalable and cost-effective manufacturing processes, facilitating the commercial viability of aerogel-based catalysts for large-scale industrial applications. Continuous innovation in synthesis methodologies, materials engineering, and manufacturing technologies has played a pivotal role in driving these advancements. New synthesis methods enable the production of aerogels with enhanced properties and improved efficiency, while advancements in materials engineering allow for the customization of aerogel-based catalysts to meet specific performance requirements. Moreover, developments in manufacturing technologies have led to increased production throughput and reduced production costs, making aerogel-based catalysts economically competitive with traditional catalyst materials. These advancements have expanded the potential applications of aerogel-based catalysts across various industries, including petrochemicals, environmental remediation, and renewable energy. As a result, aerogel-based catalysts are increasingly recognized as valuable tools for addressing key challenges in catalysis and contributing to sustainable industrial processes (Tofanica et al. 2022).

In general, aerogel-based catalysts offer a compelling combination of high surface area, tunable pore structure, uniform dispersion of active sites, thermal stability, chemical inertness, and versatility, making them highly attractive for catalytic applications across various industries. Their unique properties hold immense potential for advancing catalysis, environmental remediation, energy production, and other critical areas of research and technology development (Noman et al., 2021).

4.5 Synthesis Methods of Aerogel-Based Catalysts

Aerogel-based catalysts can be synthesized using various methods, each offering unique advantages (**Table 2**). The sol-gel method involves the formation of a sol (colloidal suspension) followed by gelation to produce the aerogel. This approach allows precise control over composition and structure. Supercritical drying utilizes supercritical fluid to remove solvent from the gel, preserving the porous structure and preventing shrinkage. Aerosol-assisted synthesis involves the generation of aerosol droplets containing precursor materials, facilitating homogeneous deposition on substrates. Template-assisted synthesis uses templates to control the pore structure and morphology of the aerogel during synthesis, enabling precise tuning of pore size and shape. These methods offer flexibility in tailoring the properties of aerogel-based catalysts for various applications in catalysis, environmental remediation, and energy conversion (Khan et al. 2023).

Table 2. Synthesis methods of aerogel-based catalysts

Synthesis Method	Description	Advantages	References
Sol-Gel Method	Involves the formation of a sol (colloidal suspension) followed by gelation to form the aerogel.	Precise control over composition and structure	Navas et al. 2021.
Supercritical Drying	Utilizes supercritical fluid to remove solvent from the gel, leaving behind the aerogel structure.	Preserves porous structure and prevents shrinkage	Franco et al. 2021.
Aerosol-Assisted Synthesis	Involves the generation of aerosol droplets containing precursor materials, followed by gelation and drying.	Allows for homogeneous deposition on substrates	Majerič & Rudolf, 2020.
Template-Assisted Synthesis	Uses templates to control the pore structure and morphology of the aerogel during synthesis.	Enables precise tuning of pore size and shape	Karamikamkar et al. 2023.

4.6 APPLICATIONS OF AEROGEL-BASED CATALYSTS

Aerogel-based catalysts have garnered attention across various industries due to their unique properties and versatile applications (**Table 3**). One prominent application is environmental remediation, where aerogels are utilized in catalytic converters to mitigate emissions from vehicles by facilitating more efficient conversion of harmful gases into less harmful substances. Their high surface area allows for enhanced catalytic activity, improving the overall effectiveness of pollution control systems. In the realm of energy production, aerogel catalysts play a crucial role in hydrogen generation processes such as steam reforming. Their enhanced catalytic activity enables more efficient conversion of feedstocks into hydrogen, a key component in various clean energy applications. Additionally, aerogel-based catalysts find applications in petrochemical processing, where they are employed in refining processes to produce fuels and chemicals. Their high thermal stability makes them suitable for harsh operating conditions encountered in such industries.

Furthermore, aerogels can be functionalized for water purification purposes, effectively removing pollutants and contaminants from water sources. Their large surface area allows for increased adsorption capacity, making them highly effective in water treatment applications. Overall, aerogel-based catalysts demonstrate immense potential across diverse sectors, offering innovative solutions for environmental sustainability and industrial efficiency (Saleh et al. 2023).

Table 3. Applications of aerogel-based catalysts

Application	Description	Benefits	References
Environmental Remediation	Aerogel-based catalysts can be utilized in catalytic converters to reduce emissions from vehicles.	High surface area for efficient catalysis	Yang et al. 2023.
Hydrogen Production	Aerogels can serve as catalysts in processes like steam reforming for hydrogen generation.	Enhanced catalytic activity	Mosińska et al. 2020.
Petrochemical Processing	Aerogel-based catalysts can be used in refining processes for the production of fuels and chemicals.	High thermal stability for harsh conditions	Zhao et al. 2023.
Water Purification	Aerogels can be functionalized to remove pollutants and contaminants from water sources.	Large surface area for adsorption	Romero-Montero et al. 2023.

4.7 EMERGING TRENDS IN AEROGEL-BASED CATALYSIS

Emerging trends in aerogel-based catalysis reflect ongoing efforts to enhance the performance, versatility, and sustainability of catalysts for various applications. These trends encompass novel synthesis methods, innovative catalyst design strategies, and the integration of biocatalysis into aerogel matrices (Obeid & Younes,2024).

Sustainable Catalyst Synthesis: As sustainability becomes increasingly important, there is a growing emphasis on developing aerogel-based catalysts using renewable precursors and environmentally friendly synthesis methods. Researchers are exploring the use of biomass-derived precursors, such as cellulose or lignin, to produce aerogels with reduced environmental impact. Additionally, eco-friendly synthesis routes, such as sol-gel processes using water-based solvents or green chemistry principles, are being investigated to minimize energy consumption and waste generation during catalyst production. These sustainable synthesis approaches not only contribute to reducing the carbon footprint of aerogel-based catalysts but also align with the principles of green chemistry (Ahmad et al. 2023).

Nanoparticle Integration: Incorporating nanoparticles into aerogel matrices represents a key trend in enhancing catalytic activity and selectivity. Nanoparticles, such as metal nanoparticles or semiconductor quantum dots, can be dispersed within the aerogel structure to create hybrid materials with synergistic properties. The high surface area and porous structure of aerogels facilitate the uniform dispersion of nanoparticles, leading to enhanced catalytic performance. Moreover, the controlled synthesis of nanoparticle-loaded aerogels allows for precise tuning of particle size, morphology, and distribution, further optimizing catalytic activity for specific reactions. These nanoparticle-integrated aerogel catalysts hold promise for applications in fields such as energy conversion, environmental remediation, and chemical synthesis (Gandla et al. 2023).

Multifunctional Catalyst Design: Another emerging trend is the development of multifunctional aerogel-based catalysts with multiple active sites or functionalities. By integrating different catalytic species or functional additives into aerogel matrices, researchers can design catalysts capable of performing complex reactions or catalyzing multiple sequential steps within a single system. For example, aerogel-supported catalysts may combine acid and metal catalytic sites to facilitate tandem reactions, such as hydrocarbon upgrading or biomass conversion. Additionally, functional additives, such as photoactive compounds or redox mediators, can be incorporated into aerogel catalysts to enable light-driven or electrochemical transformations. These multifunctional catalysts offer versatility and efficiency in catalyzing diverse reactions, making them valuable for applications requiring complex chemical transformations (Farpón et al. 2023).

Biocatalysis Integration: Integration of biocatalysis into aerogel matrices represents a promising trend in expanding the scope of aerogel-based catalysis. Enzymes immobilized on aerogel surfaces offer advantages such as enhanced stability, recyclability, and substrate specificity, making them attractive candidates for biocatalytic reactions. Aerogel matrices provide a stable and biocompatible environment for enzyme immobilization, ensuring prolonged enzyme activity under harsh reaction conditions. Furthermore, the porous structure of aerogels facilitates efficient substrate diffusion, allowing for rapid enzyme-substrate interactions and improved reaction rates. Biocatalytic aerogel-based catalysts find applications in various fields, including pharmaceuticals, food processing, and biofuel production, where they enable sustainable and environmentally friendly synthesis routes (Sandoval & Hyster, 2020).

In addition, emerging trends in aerogel-based catalysis reflect advancements in sustainable synthesis methods, nanoparticle integration, multifunctional catalyst design, and biocatalysis integration. These trends hold promise for enhancing catalytic performance, expanding the range of applications, and promoting sustainability in chemical processes. As research in aerogel-based catalysis continues to evolve, innovative approaches and novel materials are expected to drive further progress in this field, offering solutions to global challenges in energy, environment, and chemical synthesis (Trache et al. 2020; Inobeme et al. 2024).

4.8 IMPORTANCE OF AEROGELS IN CATALYSIS AND FUTURE PROSPECTS

Aerogels play a crucial role in catalysis due to their unique properties, which make them highly effective catalyst supports. The importance of aerogels in catalysis stems from their high surface area, tunable porosity, and tailored surface chemistry, all of which contribute to enhancing catalytic activity, selectivity, and stability. One of the key advantages of aerogels in catalysis is their exceptionally high surface area-to-volume ratio. This property provides a large number of active sites for catalytic reactions to occur, resulting in enhanced catalytic activity. Additionally, the tunable porosity of aerogels allows for precise control over the size and distribution of pores, facilitating efficient mass transfer of reactants and products. This is essential for improving reaction kinetics and product yields. Furthermore, aerogels offer the flexibility to tailor their surface chemistry by functionalizing them with specific groups or nanoparticles. This enables the immobilization of catalytic species, such as metal nanoparticles or enzymes, onto the aerogel surface. The presence of these catalytic species enhances the overall catalytic performance of aerogel-based catalysts, making them highly versatile for a wide range of catalytic reactions (Khan et al. 2023).

In terms of future prospects, aerogels hold significant promise for catalysis in several main areas.

Environmental Remediation: Aerogel-based catalysts demonstrate significant promise for environmental remediation applications, particularly in the degradation of pollutants in air and water. Their unique combination of high surface area and tailored surface chemistry renders them highly effective for both adsorbing and catalytically converting harmful contaminants into less toxic substances. The high surface area of aerogels provides a large number of active sites for adsorption, allowing them to efficiently capture pollutants from air and water. Additionally, their tailored surface chemistry enables selective adsorption of specific contaminants, enhancing the removal efficiency of targeted pollutants (Asghar et al. 2024).

Furthermore, aerogel-based catalysts facilitate the catalytic conversion of adsorbed pollutants into less harmful substances through various chemical reactions. For example, catalytic oxidation or reduction reactions can be employed to transform organic pollutants into carbon dioxide and water, or to convert toxic metals into less toxic forms. The presence of catalytic species on the aerogel surface enhances the reaction rates and efficiency of these processes, leading to more rapid and effective pollutant degradation. Overall, the versatility and effectiveness of aerogel-based catalysts make them promising candidates for addressing environmental pollution challenges. As research in this field continues to advance, aerogel-based catalysts are expected to play an increasingly important role in mitigating pollution and promoting environmental sustainability (Wang et al. 2023; Mathew et al. 2024c).

Sustainable Chemistry: As sustainability becomes increasingly important, aerogels present opportunities for green chemistry applications. By utilizing renewable precursors and environmentally friendly synthesis methods, aerogel-based catalysts can play a significant role in developing sustainable chemical processes with reduced environmental impact. One key advantage of aerogels is their ability to be synthesized from renewable raw materials, such as silica derived from biomass or waste sources. This reduces dependence on fossil fuels and minimizes the carbon footprint associated with aerogel production. Additionally, the synthesis of aerogels can be carried out using solvent-free or water-based processes, further reducing environmental impact and minimizing the generation of hazardous waste (Ahmad et al. 2023).

Aerogel-based catalysts offer several benefits for green chemistry applications. Their high surface area and tailored surface chemistry make them effective catalyst supports, enabling efficient conversion of feedstocks into desired products while minimizing energy consumption and waste generation. Furthermore, the recyclability and durability of aerogel-based catalysts contribute to the sustainability of chemical processes by reducing the need for frequent replacement and disposal. Furthermore, aerogels hold great potential for advancing green chemistry initiatives and promoting the development of sustainable chemical processes. By harnessing the unique properties of aerogels and adopting environmentally friendly synthesis methods, researchers can contribute to the transition towards a more sustainable and environmentally conscious chemical industry (Bari & Jeong, 2024).

Energy Conversion: Aerogels have emerged as promising materials in energy conversion applications, including fuel cells and photocatalytic water splitting for hydrogen production. Their exceptional properties, such as high surface area and efficient mass transport, render them well-suited for supporting catalysts in these systems, thereby enhancing efficiency and performance. In fuel cells, aerogels serve as excellent catalyst supports due to their large surface area, which allows for enhanced interaction between the catalyst and reactants. This facilitates more efficient electrochemical reactions, leading to improved fuel cell performance and higher energy conversion efficiencies (TinocoNavarro & Jaroslav, 2023; Mathew et al. 2024d).

Similarly, in photocatalytic water splitting for hydrogen production, aerogels play a crucial role as catalyst supports. Their high surface area provides ample active sites for photocatalytic reactions, while their efficient mass transport properties ensure rapid diffusion of reactants and products. This results in improved rates of hydrogen production and overall process efficiency. However, the use of aerogels in energy conversion applications holds great promise for advancing renewable energy technologies. As research in this field continues to progress, aerogels are expected to play an increasingly important role in improving the efficiency and sustainability of energy conversion processes, contributing to the transition towards a cleaner and more sustainable energy future (Kubovics et al. 2022).

Biocatalysis: Aerogels are increasingly recognized for their potential in biocatalytic applications, particularly in the field of biocatalysis, where enzymes immobilized on aerogel surfaces catalyze specific chemical reactions. This approach presents several advantages, including enhanced enzyme stability and recyclability, which are highly advantageous for industries such as pharmaceuticals and biofuels. Enzymes immobilized on aerogel surfaces benefit from the high surface area and porous structure of aerogels, which provide ample space for enzyme attachment and facilitate efficient substrate binding. Additionally, the inert nature of aerogels helps to stabilize the enzymes, protecting them from denaturation and degradation under harsh reaction conditions. This enhanced stability extends the lifespan of the biocatalysts, resulting in longer enzyme activity and increased productivity (Efremenko et al. 2023).

Furthermore, aerogel-based biocatalysts offer the advantage of recyclability, allowing for the recovery and reuse of enzymes in multiple reaction cycles. This not only reduces the overall cost of enzyme production but also minimizes waste generation and environmental impact. Overall, the use of aerogels in biocatalytic applications holds great promise for improving process efficiency and sustainability in industries such as pharmaceuticals and biofuels. As research in this area continues to advance, aerogel-based biocatalysts are expected to play an increasingly important role in driving innovation and addressing challenges in biocatalysis (Li et al. 2019).

Advanced Materials: Recent advancements in aerogel synthesis and functionalization techniques have paved the way for the development of advanced catalytic materials with tailored properties. By incorporating nanoparticles, quantum dots, or other functional additives into aerogel matrices, researchers can engineer hybrid materials that exhibit enhanced catalytic performance and multifunctionality. One approach involves the integration of metal nanoparticles, such as platinum or palladium, into aerogel structures to create efficient catalysts for various chemical reactions. These metal nanoparticles can serve as active sites for catalyzing reactions like hydrogenation, oxidation, or carbon dioxide reduction. Additionally, the high surface area and porous nature of aerogels provide ample opportunities for maximizing the exposure of active sites and promoting catalytic efficiency (Bari et al. 2024).

Another strategy is the incorporation of semiconductor quantum dots, such as titanium dioxide or zinc oxide, into aerogel frameworks for photocatalytic applications. These quantum dots can harness solar energy to drive catalytic reactions, such as water splitting or pollutant degradation, under ambient conditions. By coupling the unique properties of quantum dots with the structural advantages of aerogels, researchers can design highly efficient and environmentally friendly photocatalysts. Generally, these advancements in aerogel-based catalytic materials offer promising prospects for addressing challenges in energy conversion, environmental remediation, and sustainable chemistry. By tailoring the properties of aerogels through innovative synthesis and functionalization approaches, researchers can unlock new opportunities for catalysis and contribute to the development of next-generation materials with enhanced performance and versatility (Li et al., 2019).

Overall, the future of aerogels in catalysis looks bright, with ongoing research focused on developing innovative materials and processes for addressing global challenges in energy, environment, and sustainability. As scientists continue to explore the diverse capabilities of aerogels and push the boundaries of catalytic science, aerogel-based catalysts are poised to play an increasingly important role in driving technological advancements and meeting the demands of a rapidly changing world (Horvat & Maver, 2024).

4.9 CONCLUSION

Generally, aerogels exhibit immense potential as catalyst supports across various catalytic applications due to their unique combination of properties. Their high surface area, tunable porosity, and tailored surface chemistry make them ideal candidates for enhancing catalytic activity, selectivity, and stability. Aerogel-based catalysts have demonstrated superior performance compared to traditional catalysts in heterogeneous, photocatalytic, and enzymatic reactions. The efficient mass transfer and active site accessibility afforded by aerogels contribute to improved reaction kinetics and product yields. Furthermore, the versatility of aerogels allows for the immobilization of various catalytic species, enabling the design of tailored catalysts for specific reactions. Recent advances in aerogel synthesis and functionalization methods have expanded their applications in environmental remediation, chemical synthesis, and energy conversion. Looking ahead, continued research efforts in aerogel-based catalysis hold promise for addressing global challenges and advancing sustainable technologies. Further exploration of novel aerogel compositions, synthesis techniques, and catalytic functionalities will unlock new opportunities for catalytic applications, driving innovation in diverse fields. Overall, aerogels represent a valuable platform for catalysis, offering tailored solutions for a wide range of industrial and environmental challenges.

REFERENCES

Abbas, Q., Shinde, P. A., Abdelkareem, M. A., Alami, A. H., Mirzaeian, M., Yadav, A., & Olabi, A. G. (2022). Graphene Synthesis Techniques and Environmental Applications. *Materials (Basel), 15*(21), 7804. DOI: 10.3390/ma15217804 PMID: 36363396

Adetunji, C. O., Mathew, J. T., Inobeme, A., Taiwo, O., Olaniyan, O., Shakira, G., & Wadazani, D.(2023).Application of Biosensors for Detection and Monitoring of Water Quality.In book: Organic and Inorganic Materials Based Sensors, John Wiley & Sons, Inc., 1065-1077, DOI: 10.3390/ma15217804 PMID: 36363396

Ahmad, A., & Kamaruddin, M. A., H P S, A. K., Yahya, E. B., Muhammad, S., Rizal, S., Ahmad, M. I., Surya, I., & Abdullah, C. K. (. (2023). Recent Advances in Nanocellulose Aerogels for Efficient Heavy Metal and Dye Removal. *Gels (Basel, Switzerland), 9*(5), 416. DOI: 10.3390/gels9050416 PMID: 37233007

Anžlovar, A., & Žagar, E. (2022). Cellulose Structures as a Support or Template for Inorganic Nanostructures and Their Assemblies. *Nanomaterials (Basel, Switzerland), 12*(11), 1837. DOI: 10.3390/nano12111837 PMID: 35683693

Arshad, A., Roy, A., Mallick, T. K., & Tahir, A. A. (2023). Shape-Stabilized PEGylated Silica Aerogel-Composite as an Energy Saving Building Material. *Industrial & Engineering Chemistry Research, 62*(47), 20236–20250. DOI: 10.1021/acs.iecr.3c02373 PMID: 38045733

Asghar, N., Hussain, A., Nguyen, D. A., Ali, S., Hussain, I., Junejo, A., & Ali, A. (2024). Advancement in nanomaterials for environmental pollutants remediation: A systematic review on bibliometrics analysis, material types, synthesis pathways, and related mechanisms. *Journal of Nanobiotechnology, 22*(1), 26. DOI: 10.1186/s12951-023-02151-3 PMID: 38200605

Bakhori, N. M., Ismail, Z., Hassan, M. Z., & Dolah, R. (2023). Emerging Trends in Nanotechnology: Aerogel-Based Materials for Biomedical Applications. *Nanomaterials (Basel, Switzerland), 13*(6), 1063. DOI: 10.3390/nano13061063 PMID: 36985957

Bari, G. A. K. M. R., & Jeong, J. H. (2024). Comprehensive Insights and Advancements in Gel Catalysts for Electrochemical Energy Conversion. *Gels (Basel, Switzerland)*, *10*(1), 63. DOI: 10.3390/gels10010063 PMID: 38247786

Budtova, T., Aguilera, D. A., Beluns, S., Berglund, L., Chartier, C., Espinosa, E., Gaidukovs, S., Klimek-Kopyra, A., Kmita, A., Lachowicz, D., Liebner, F., Platnieks, O., Rodríguez, A., Navarro, L. K. T., Zou, F., & Buwalda, S. J. (2020). Biorefinery Approach for Aerogels. *Polymers*, *12*(12), 2779. DOI: 10.3390/polym12122779 PMID: 33255498

Efremenko, E., Stepanov, N., Senko, O., Maslova, O., Lyagin, I., & Aslanli, A. (2023). Progressive Biocatalysts for the Treatment of Aqueous Systems Containing Pharmaceutical Pollutants. *Life (Basel, Switzerland)*, *13*(3), 841. DOI: 10.3390/life13030841 PMID: 36983996

Farpón, M. G., Peláez, R., Recio, V., Atakan, B., Zaldo, C., & Prieto, G. (2023). Multifunctional materials for catalyst-specific heating and thermometry in tandem catalysis. *Journal of Materials Chemistry. A, Materials for Energy and Sustainability*, *11*(37), 19854–19859. DOI: 10.1039/D3TA03654E PMID: 38013847

Franco, P., Cardea, S., Tabernero, A., & De Marco, I. (2021). Porous Aerogels and Adsorption of Pollutants from Water and Air: A Review. *Molecules (Basel, Switzerland)*, *26*(15), 4440. DOI: 10.3390/molecules26154440 PMID: 34361593

Gandla, K., Kumar, K. P., Rajasulochana, P., Charde, M. S., Rana, R., Singh, L. P., Haque, M. A., Bakshi, V., Siddiqui, F. A., Khan, S. L., & Ganguly, S. (2023). Fluorescent-Nanoparticle-Impregnated Nanocomposite Polymeric Gels for Biosensing and Drug Delivery Applications. *Gels (Basel, Switzerland)*, *9*(8), 669. DOI: 10.3390/gels9080669 PMID: 37623124

Gao, J., Ma, R., Feng, L., Liu, Y., Jackstell, R., Jagadeesh, R. V., & Beller, M. (2021). Ambient Hydrogenation and Deuteration of Alkenes Using a Nanostructured Ni-Core-Shell Catalyst. *Angewandte Chemie International Edition*, *60*(34), 18591–18598. DOI: 10.1002/anie.202105492 PMID: 34076934

Guengerich, F. P., & Yoshimoto, F. K. (2018). Formation and Cleavage of C-C Bonds by Enzymatic Oxidation-Reduction Reactions. *Chemical Reviews*, *118*(14), 6573–6655. DOI: 10.1021/acs.chemrev.8b00031 PMID: 29932643

Hayes, G., Laurel, M., MacKinnon, D., Zhao, T., Houck, H. A., & Becer, C. R. (2023). Polymers without Petrochemicals: Sustainable Routes to Conventional Monomers. *Chemical Reviews*, *123*(5), 2609–2734. DOI: 10.1021/acs.chemrev.2c00354 PMID: 36227737

Horvat, G., & Maver, U. (2024). Editorial on the Special Issue Entitled "Recent Advances in Aerogels". *Gels (Basel, Switzerland)*, *10*(3), 154. DOI: 10.3390/gels10030154 PMID: 38534572

Inobeme, A., Adetunji, C. O., Mathew, J T., Ajai, A. I., Inobeme, J., Bamigboye, M. O., Onyeaku, S., Maliki, M., Eziukwu, C. &Tawa, K. (2023). Nanotechnology for Bioremediation of Heavy Metals. In book: Microbial Technologies in Industrial Wastewater Treatment. Springer Nature Singapore, 19-30. .DOI: 10.1007/978-981-99-2435-6_2

Inobeme, A., Mathew, J.T., Adetunji, C.O., Agbugui, M.O., Osarenren, E.J., Oti, C., Ngonso, B.F., Inobeme, J., Oyedolapo, M.B., Bernard, E. &Jibrin, N.A. (2024). Biocomposites with graphene derivatives. In book: Advances in Biocomposites and their Applications, Elsevier Inc., 149-166. DOI: DOI: 10.1016/B978-0-443-19074-2.00005-8

Inobeme, A., Shahnawaz, M., Adetunji, C. O., Mathew, J. T., & Adetuyi, B. O. Popoola, Oluwabukola Atinuke; Olaitan, Falana Yetunde; Akinbo, O., Kolawole, O. M., Oyewole, O. A., Eniola, K.I.T., Yerima, M. B. (2024). Primary Microplastic: Source, Uses, Transportation from Land to Marine Environment. In book: Microplastic Pollution. Springer, Singapore, 1, 23–33. DOI: DOI: 10.1007/978-981-99-8357-5_2

Jin, R., Zhou, Z., Liu, J., Shi, B., Zhou, N., Wang, X., Jia, X., Guo, D., & Xu, B. (2023). Aerogels for Thermal Protection and Their Application in Aerospace. *Gels (Basel, Switzerland)*, *9*(8), 606. DOI: 10.3390/gels9080606 PMID: 37623061

Karamikamkar, S., Yalcintas, E. P., Haghniaz, R., de Barros, N. R., Mecwan, M., Nasiri, R., Davoodi, E., Nasrollahi, F., Erdem, A., Kang, H., Lee, J., Zhu, Y., Ahadian, S., Jucaud, V., Maleki, H., Dokmeci, M. R., Kim, H. J., &Khademhosseini, A. (2023). Aerogel-Based Biomaterials for Biomedical Applications: From Fabrication Methods to Disease-Targeting Applications. Advanced science (Weinheim, Baden-Wurttemberg, Germany), 10(23), e2204681. DOI: 10.1002/advs.202204681

Khan, N. R., Sharmin, T., & Bin Rashid, A. (2023). Exploring the Versatility of Aerogels: Broad Applications in Biomedical Engineering, Astronautics, Energy Storage, Biosensing, and Current Progress. *Heliyon*, *10*(1), e23102. DOI: 10.1016/j.heliyon.2023.e23102 PMID: 38163169

Khandelwal, K., Boahene, P., Nanda, S., & Dalai, A. K. (2023). A Review of the Design and Performance of Catalysts for Hydrothermal Gasification of Biomass to Produce Hydrogen-Rich Gas Fuel. *Molecules (Basel, Switzerland)*, *28*(13), 5137. DOI: 10.3390/molecules28135137 PMID: 37446799

Kubovics, M., Silva, C. G., López-Periago, A. M., Faria, J. L., & Domingo, C. (2022). Photocatalytic Hydrogen Production using Porous 3D Graphene-Based Aerogels Supporting Pt/TiO2 Nanoparticles. *Gels (Basel, Switzerland)*, *8*(11), 719. DOI: 10.3390/gels8110719 PMID: 36354627

Lermontov, S. A., Vlasenko, N. E., Sipyagina, N. A., Malkova, A. N., Gozhikova, I. O., Baranchikov, A. E., & Knerelman, E. I. (2023). Highly Porous Para-Aramid Aerogel as a Heterogeneous Catalyst for Selective Hydrogenation of Unsaturated Organic Compounds. *Polymers*, *15*(15), 3206. DOI: 10.3390/polym15153206 PMID: 37571100

Li, L. J., Xia, W. J., Ma, G. P., Chen, Y. L., & Ma, Y. Y. (2019). A study on the enzymatic properties and reuse of cellulaseimmobilized with carbon nanotubes and sodium alginate. *AMB Express*, *9*(1), 112. DOI: 10.1186/s13568-019-0835-0 PMID: 31332555

Liu, Z., Xu, B., Jiang, Y. J., Zhou, Y., Sun, X., Wang, Y., & Zhu, W. (2023). Photocatalytic Conversion of Methane: Current State of the Art, Challenges, and Future Perspectives. *ACS Environmental Au*, *3*(5), 252–276. DOI: 10.1021/acsenvironau.3c00002 PMID: 37743954

Luo, Y., Mei, Y., Xu, Y., & Huang, K. (2023). Hyper-Crosslinked Porous Organic Nanomaterials: Structure-Oriented Design and Catalytic Applications. *Nanomaterials (Basel, Switzerland)*, *13*(18), 2514. DOI: 10.3390/nano13182514 PMID: 37764543

Maio, A., Pibiri, I., Morreale, M., Mantia, F. P., & Scaffaro, R. (2021). An Overview of Functionalized GrapheneNanomaterials for Advanced Applications. *Nanomaterials (Basel, Switzerland)*, *11*(7), 1717. DOI: 10.3390/nano11071717 PMID: 34209928

Majerič, P., & Rudolf, R. (2020). Advances in Ultrasonic Spray Pyrolysis Processing of Noble Metal Nanoparticles-Review. *Materials (Basel)*, *13*(16), 3485. DOI: 10.3390/ma13163485 PMID: 32784637

Mathew, J. T., Inobeme, A., Abdulkadir, A., Etsuyankpa, M. B., Shaba, E. Y., Musah, M., Azeh, Y., Adetunji, C. O., Ibrahim, A. M., Tanko, E., & Mamman, A. (2024). a. Properties and Characteristics of Nanocoatings for Medicinal Applications. In book: Sustainable Approach to Protective Nanocoatings. IGI Global; publishing Tomorrow's. *Advances in Chemical and Materials Engineering*, •••, 203–227. DOI: 10.4018/979-8-3693-3136-1.ch007

Mathew, J. T., Inobeme, A., Adetuyi, B. O., Adetunji, C. O., Popoola, O. A., Olaitan, F. Y., Akinbo, O., Shahnawaz, M., Oyewole, O. A., Eniola, K. I. T., & Yerima, B. M. (2024)c. General Introduction of Microplastic: Uses, Types, and Generation. In book: Microplastic Pollution. Springer, Singapore, 1, 3-21. DOI: DOI: 10.1007/978-981-99-8357-5_1

Mathew, J. T., Inobeme, A., Etsuyankpa, M. B., Adetunji, C. O., Shaba, E. Y., Musah, M., Azeh, Y., Jiya, M. J., Tanko, E., Muhammad, A. I., Mamman, A. & Maurice, J. (2024)d. Applications of Diatom for Treatment of Industrial Effluents: Waste to Wealth Approach. . In book: Marine Biomass: Biorefinery, Bioproducts and Environmental Bioremediation. Walter de Gruyter GmbH & Co KG, 1-20, DOI: DOI: 10.1515/9783111353951-001

Mathew, J. T., Inobeme, A., Musah, M., Azeh, Y., Abdullahi, A., Shaba, E. Y., Salihu, A. M., Muhammad, E. B., Josiah, J. G., Jibrin, N. A., Ismail, H., Muhammad, A. I., Maurice, J., Mamman, A., & Ndamitso, M. M. (2024). b. A Critical Review of Green Approach on Wastewater Treatment Strategies. *Journal of Applied Science & Environmental Management*, *28*(2), 363–391. DOI: 10.4314/jasem.v28i2.9

Mathew, J. T., Musah, M., Azeh, Y., & Muhammed, M. (2023). a. Kinetic Study of Heavy Metals Removal from Pharmaceutical Wastewater Using Geopolymer/Fe_3O_4Nanocomposite. *Bima Journal of Science and Technology*, *7*(4), 152–163. DOI: 10.56892/bima.v7i4.539

Mathew, J. T., Musah, M., Azeh, Y., & Muhammed, M. (2023). b. Adsorptive Removal of Selected Toxic Metals from Pharmaceutical Wastewater using Fe_3O_4/ZnONanocomposite. *Dutse Journal of Pure and Applied Sciences*, *9*(4a), 236–248. DOI: 10.4314/dujopas.v9i4a.22

McKay, A. I., Bukvic, A. J., Tegner, B. E., Burnage, A. L., Martı Nez-Martı Nez, A. J., Rees, N. H., Macgregor, S. A., & Weller, A. S. (2019). Room Temperature Acceptorless Alkane Dehydrogenation from Molecular σ-Alkane Complexes. *Journal of the American Chemical Society*, *141*(29), 11700–11712. DOI: 10.1021/jacs.9b05577 PMID: 31246012

Mosińska, M., Szynkowska-Jóźwik, M. I., & Mierczyński, P. (2020). Catalysts for Hydrogen Generation via Oxy-Steam Reforming of Methanol Process. *Materials (Basel)*, *13*(24), 5601. DOI: 10.3390/ma13245601 PMID: 33302526

Navas, D., Fuentes, S., Castro-Alvarez, A., & Chavez-Angel, E. (2021). Review on Sol-Gel Synthesis of Perovskite and Oxide Nanomaterials. *Gels (Basel, Switzerland)*, *7*(4), 275. DOI: 10.3390/gels7040275 PMID: 34940335

Nghiem, T. L., Coban, D., Tjaberings, S., & Gröschel, A. H. (2020). Recent Advances in the Synthesis and Application of Polymer Compartments for Catalysis. *Polymers*, *12*(10), 2190. DOI: 10.3390/polym12102190 PMID: 32987965

Noman, M. T., Amor, N., Ali, A., Petrik, S., Coufal, R., Adach, K., & Fijalkowski, M. (2021). Aerogels for Biomedical, Energy and Sensing Applications. *Gels (Basel, Switzerland)*, *7*(4), 264. DOI: 10.3390/gels7040264 PMID: 34940324

Obeid, E., & Younes, K. (2024). Uncovering Key Factors in Graphene Aerogel-Based Electrocatalysts for Sustainable Hydrogen Production: An Unsupervised Machine Learning Approach. *Gels (Basel, Switzerland)*, *10*(1), 57. DOI: 10.3390/gels10010057 PMID: 38247780

Potter, M. E., Le Brocq, J. J. M., Oakley, A. E., McShane, E. B., Vandegehuchte, B. D., & Raja, R. (2020). Butane Isomerization as a Diagnostic Tool in the Rational Design of Solid Acid Catalysts. *Catalysts*, *10*(9), 1099. DOI: 10.3390/catal10091099

Romero-Montero, A., Valencia-Bermúdez, J. L., Rosas-Meléndez, S. A., Núñez-Tapia, I., Piña-Barba, M. C., Leyva-Gómez, G., & Del Prado-Audelo, M. L. (2023). Biopolymeric Fibrous Aerogels: The Sustainable Alternative for Water Remediation. *Polymers*, *15*(2), 262. DOI: 10.3390/polym15020262 PMID: 36679143

Saleh, W.M., Ahmad, M.I., & Yahya, E.B., & H.P.S., A.K. (. (2023). Nanostructured Bioaerogels as a Potential Solution for Particulate Matter Pollution. *Gels (Basel, Switzerland)*, *9*, 575. DOI: 10.3390/gels9070575 PMID: 37504454

Sandoval, B. A., & Hyster, T. K. (2020). Emerging strategies for expanding the toolbox of enzymes in biocatalysis. *Current Opinion in Chemical Biology*, *55*, 45–51. DOI: 10.1016/j.cbpa.2019.12.006 PMID: 31935627

Shan, J., Shan, Y., Zou, C., Hong, Y., Liu, J., & Guo, X. (2024). Cost-Effective Preparation of Hydrophobic and Thermal-Insulating Silica Aerogels. *Nanomaterials (Basel, Switzerland)*, *14*(1), 119. DOI: 10.3390/nano14010119 PMID: 38202574

Sozcu, S., Venkataraman, M., Wiener, J., Tomkova, B., Militky, J., & Mahmood, A. (2023). Incorporation of Cellulose-Based Aerogels into Textile Structures. *Materials (Basel)*, *17*(1), 27. DOI: 10.3390/ma17010027 PMID: 38203881

Sun, B., Wang, B., & Wang, R. (2023). Advances in Emerging Catalytic Materials for the Conversion of Carbon Dioxide. *Materials (Basel), 16*(23), 7309. DOI: 10.3390/ma16237309 PMID: 38068053

Tebes-Stevens, C., Patel, J. M., Jones, W. J., & Weber, E. J. (2017). Prediction of Hydrolysis Products of Organic Chemicals under Environmental pH Conditions. *Environmental Science & Technology, 51*(9), 5008–5016. DOI: 10.1021/acs.est.6b05412 PMID: 28430419

Tinoco Navarro, L. K., & Jaroslav, C. (2023). Enhancing Photocatalytic Properties of TiO2 Photocatalyst and Heterojunctions: A Comprehensive Review of the Impact of Biphasic Systems in Aerogels and Xerogels Synthesis, Methods, and Mechanisms for Environmental Applications. *Gels (Basel, Switzerland), 9*(12), 976. DOI: 10.3390/gels9120976 PMID: 38131962

Tofanica, B. M., Belosinschi, D., & Volf, I. (2022). Gels, Aerogels and Hydrogels: A Challenge for the Cellulose-Based Product Industries. *Gels (Basel, Switzerland), 8*(8), 497. DOI: 10.3390/gels8080497 PMID: 36005098

Trache, D., Tarchoun, A. F., Derradji, M., Hamidon, T. S., Masruchin, N., Brosse, N., & Hussin, M. H. (2020). Nanocellulose: From Fundamentals to Advanced Applications. *Frontiers in Chemistry, 8*, 392. DOI: 10.3389/fchem.2020.00392 PMID: 32435633

Wang, H., Wang, L., & Xiao, F. S. (2020). Metal@Zeolite Hybrid Materials for Catalysis. *ACS Central Science, 6*(10), 1685–1697. DOI: 10.1021/acscentsci.0c01130 PMID: 33145408

Wang, X., Liu, T., Li, H., Han, C., Su, P., Ta, N., Jiang, S. P., Kong, B., Liu, J., & Huang, Z. (2023). Balancing Mass Transfer and Active Sites to Improve Electrocatalytic Oxygen Reduction by B,NCodoped C Nanoreactors. *Nano Letters, 23*(11), 4699–4707. DOI: 10.1021/acs.nanolett.3c00202 PMID: 36951377

Yang, Z., Zhou, S., Feng, X., Wang, N., Ola, O., & Zhu, Y. (2023). Recent Progress in Multifunctional Graphene-Based Nanocomposites for Photocatalysis and Electrocatalysis Application. *Nanomaterials (Basel, Switzerland), 13*(13), 2028. DOI: 10.3390/nano13132028 PMID: 37446544

Zhang, L., Lei, Y., He, P., Wu, H., Guo, L., & Wei, G. (2022). Carbon Material-Based Aerogels for Gas Adsorption: Fabrication, Structure Design, Functional Tailoring, and Applications. *Nanomaterials (Basel, Switzerland)*, *12*(18), 3172. DOI: 10.3390/nano12183172 PMID: 36144967

Zhao, J., Yuan, X., Wu, X., Liu, L., Guo, H., Xu, K., Zhang, L., & Du, G. (2023). Preparation of Nanocellulose-Based Aerogel and Its Research Progress in Wastewater Treatment. *Molecules (Basel, Switzerland)*, *28*(8), 3541. DOI: 10.3390/molecules28083541 PMID: 37110772

Chapter 7

Environmental and Health Impact Assessment on Aerogel Application

John Tsado Mathew
https://orcid.org/0000-0001-9514-5330
Department of Chemistry, Ibrahimm Badamasi Babangida Lapai, Nigeria

Jonathan Hussaini
https://orcid.org/0009-0004-9216-8691
Department of Chemistry, Ibrahim Badamasi Babangida University, Lapai, Nigeria

Yakubu Azeh
https://orcid.org/0000-0002-3818-674X
Department of Chemistry, Ibrahim Badamasi Babangida University, Lapai, Nigeria

Monday Musah
https://orcid.org/0000-0002-3224-1664
Department of Chemistry, Ibrahim Badamasi Babangida University, Lapai, Nigeria

Elijah Yanda Shaba
Department of Chemistry, Federal University of Technology, Minna, Nigeria

Ezekiel Tanko
Department of Chemistry, Federal University of Technology, Minna, Nigeria

Aishetu Ibrahim Muhammad
Department of Chemistry, Ibrahim Badamasi Babangida University, Lapai, Nigeria

DOI: 10.4018/979-8-3693-5971-6.ch007

Muhammad Bini Etsuyankpa
Department of Chemistry, Federal University, Lafia, Nigeria

Amos Mamman
Department of Chemistry, Ibrahim Badamasi Babangida University, Lapai, Nigeria

Tanko Safiyanu Musa
Department of Chemistry, Federal University, Lafia, Nigeria

ABSTRACT

This study evaluates the environmental and health impacts of aerogels, known for their low density, high porosity, and excellent thermal insulation. It involves a comprehensive review of literature and experimental analysis to assess factors like aerogel composition, production methods, usage scenarios, and disposal. Key environmental considerations include energy consumption, greenhouse gas emissions, and ecological toxicity. Health impacts are examined through potential exposure pathways, toxicological effects, and occupational hazards related to aerogel handling. The goal is to guide sustainable aerogel utilization by identifying and mitigating potential adverse effects on both ecosystems and human health, thereby supporting informed decision-making for safer and more responsible use of these materials.

12.1 INTRODUCTION

Aerogels, colloquially termed "frozen smoke," are a fascinating class of materials renowned for their exceptional properties, including ultralow density, high porosity, and remarkable thermal insulation capabilities. Originally developed in the 1930s by Samuel Kistler, aerogels have since evolved into a versatile material with applications spanning various industries, from aerospace and construction to electronics and healthcare. Their unique characteristics make them highly desirable for insulation, lightweight structural components, catalyst supports, and more. However, the widespread adoption of aerogels necessitates a comprehensive understanding of their environmental and health impacts (Noman et al., 2021).

The growing interest in aerogels is driven by their potential to address pressing challenges such as energy efficiency, sustainability, and resource conservation. As industries seek innovative solutions to enhance performance while minimizing environmental footprint, aerogels emerge as a promising candidate due to their lightweight nature and superior insulating properties. For instance, in the construction sector, aerogel-based insulation materials offer significant energy savings by reducing heat transfer through walls and roofs, thereby lowering heating and cooling requirements in buildings. Similarly, in transportation applications, aerogels can contribute to lightweighting efforts, improving fuel efficiency and reducing emissions (Koehle et al., 2023).

Despite their numerous advantages, the widespread adoption of aerogels raises concerns regarding their environmental and health implications throughout their lifecycle (figure 1). The production processes involved in synthesizing aerogels often require energy-intensive procedures and chemical precursors, which may contribute to greenhouse gas emissions and environmental pollution if not managed properly. Additionally, the disposal of aerogel-based products at the end of their lifespan presents challenges, as traditional waste management practices may not adequately address their unique properties (Abdul-Khalil et al., 2023).

Furthermore, the potential health risks associated with aerogel exposure, particularly during manufacturing, handling, and disposal, warrant careful examination. While aerogels are generally considered safe when properly encapsulated or handled, concerns arise regarding the inhalation of aerogel particles, which may pose respiratory hazards. Occupational exposure to aerogel dust and fibers may result in adverse health effects such as respiratory irritation, lung inflammation, and fibrosis. Therefore, a thorough assessment of aerogel's health impacts is crucial to ensuring the safety of workers and consumers alike (Keller et al., 2021).

This study aims to conduct a comprehensive environmental and health impact assessment of aerogel application across various sectors. By examining the entire lifecycle of aerogel-based products, from production to disposal, we seek to identify potential environmental hotspots and health hazards associated with their utilization. Through a combination of literature review, experimental analysis, and modeling, we aim to quantify the environmental burdens and health risks associated with aerogel application, while also exploring strategies to mitigate these impacts (Gatou et al., 2023).

The assessment will focus on key aspects such as energy consumption, greenhouse gas emissions, ecological toxicity, exposure pathways, toxicological effects, and occupational hazards. By systematically evaluating these factors, we aim to provide valuable insights into the sustainability and safety implications of aerogel utilization. Additionally, the study will highlight opportunities for innovation and improvement in aerogel manufacturing processes, product design, and end-of-life management practices to minimize adverse impacts on the environment and human health (Ankley et al., 2021).

Figure 1. Schematic representation of chapter

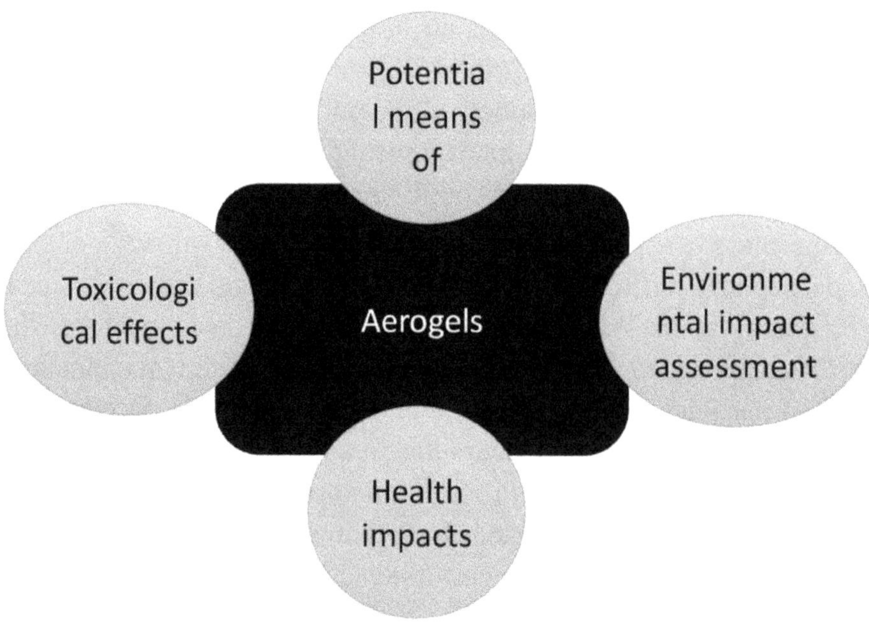

In general, while aerogels offer significant potential for enhancing performance and sustainability across various applications, their widespread adoption must be accompanied by a thorough understanding of their environmental and health implications. Through this comprehensive assessment, we aim to inform decision-makers, industry stakeholders, and researchers about the challenges and opportunities associated with aerogel application, ultimately guiding the development of more sustainable and safer practices in aerogel utilization (Vrabič-Brodnjak, 2024).

12.2 DIFFERENT TYPES OF AEROGELS AND THEIR APPLICATIONS

Aerogels come in various types, each with unique compositions and properties tailored for specific applications (**Table 1**). Silica aerogels, composed primarily of silicon dioxide, boast ultralow density, high porosity, and exceptional thermal insulation properties. These qualities make them ideal for insulation in buildings and pipelines, thermal protection in aerospace, oil spill cleanup in the oil and gas industry, lightweighting in automotive applications, and thermal apparel in clothing. Additionally, their transparency enables applications in environmental remediation, thermal management in electronics, and drug delivery in biomedical settings. Carbon aerogels, constructed from carbon-based materials like carbon fibers, graphene, or carbon nanotubes, exhibit low density, high surface area, and excellent electrical conductivity. These attributes render them suitable for energy storage in supercapacitors, structural components in aerospace, water purification through adsorption, gas detection in sensors, and tissue engineering in biomedical applications (Ghica et al., 2023).

Polymer aerogels, comprising a polymer network, offer lightweight, flexibility, thermal stability, and biocompatibility. Their tunable properties make them suitable for insulation in packaging, sportswear in clothing, air and water filtration, energy storage in batteries, and noise insulation in automotive and aerospace industries. Metal aerogels, featuring metallic nanoparticles, possess high surface area, porous structure, lightweight, and conductivity. They find applications in catalysis for chemical reactions, energy storage in batteries, thermal protection in aerospace, electromagnetic interference (EMI) shielding in electronics, and air and water filtration (Paraskevopoulou et al., 2019).

Table 1. Different types of aerogels and their applications

Aerogel Type	Composition	Properties	Applications	References
Carbon Aerogel	Carbon-based materials (carbon fibers, graphene, carbon nanotubes)	Low density High surface area Excellent electrical conductivity	Energy storage (supercapacitors) Aerospace (structural components) Water purification (adsorption) Sensors (gas detection) Biomedical (tissue engineering)	Khan et al., 2023
Silica Aerogel	Silicon dioxide	Ultralow density High porosity Excellent thermal insulation Transparency	Insulation (buildings, pipelines) Aerospace (thermal protection) Oil and gas (oil spill cleanup) Automotive (lightweighting) Clothing (thermal apparel) Environmental remediation Electronics (thermal management) Biomedical (drug delivery)	Jin et al., 2023
Polymer Aerogel	Polymer network	Lightweight Flexibility Thermal stability Biocompatibility Tunable properties	Insulation (packaging) Clothing (sportswear) Filtration (air and water) Energy storage (batteries) Automotive (noise insulation) Aerospace (structural components)	Azeh et al. 2023; Wang et al., 2023
Metal Aerogel	Metallic nanoparticles (gold, silver, copper, nickel)	High surface area Porous structure Lightweight Conductivity	Catalysis (chemical reactions) Energy storage (batteries) Aerospace (thermal protection) Electronics (EMI shielding) Filtration (air and water)	Mathew et al., 2023a; Nan et al., 2023

12.3 ENVIRONMENTAL IMPACT ASSESSMENT ON AEROGEL APPLICATION

Aerogels, with their exceptional properties such as ultralow density, high porosity, and superb thermal insulation capabilities, have garnered considerable attention for various applications. However, the widespread adoption of aerogels raises concerns regarding their environmental impact throughout their lifecycle. This assessment aims to comprehensively evaluate the environmental implications of aerogel application across different sectors (Sozcu et al., 2023).

12.3.1 Main Environmental Considerations

The production of aerogels typically involves energy-intensive processes and the use of chemical precursors, leading to potential environmental impacts (figure 2). Energy consumption during aerogel synthesis, particularly in the supercritical drying stage, contributes significantly to the overall environmental footprint. Additionally, the production of aerogel precursors may involve resource extraction, chemical synthesis, and manufacturing processes, which can result in emissions of greenhouse gases, air pollutants, and wastewater discharge (Vrabič-Brodnjak, 2024).

Energy Consumption: The energy-intensive nature of aerogel production, particularly in the supercritical drying process, leads to substantial energy consumption and associated greenhouse gas emissions. Supercritical drying, a critical step in aerogel production, requires significant amounts of energy to remove the solvent from the gel matrix without causing structural collapse. To address this challenge, strategies aimed at improving energy efficiency are essential. This includes optimizing process parameters, such as temperature and pressure, to minimize energy input while maintaining product quality. Additionally, investing in advanced drying technologies, such as microwave or infrared drying, can enhance energy efficiency and reduce greenhouse gas emissions. Furthermore, transitioning to renewable energy sources can significantly mitigate the environmental impact of aerogel production. By harnessing solar, wind, or hydroelectric power for manufacturing operations, aerogel producers can reduce reliance on fossil fuels and decrease their carbon footprint. Moreover, implementing energy management systems and adopting energy-saving measures throughout the production process can further improve energy efficiency and minimize environmental impact. Overall, a concerted effort to enhance energy efficiency and utilize renewable energy sources is essential for mitigating the environmental footprint of aerogel production (Lee et al., 2023; Xue et al., 2023).

Greenhouse Gas Emissions: The production of aerogel precursors and the energy-intensive processes involved emit significant amounts of greenhouse gases, including carbon dioxide, methane, and nitrous oxide, which contribute to climate change. To mitigate these emissions, implementing sustainable manufacturing practices is crucial. This includes optimizing production processes to minimize energy consumption, reducing waste generation, and improving resource efficiency. Additionally, transitioning to low-carbon energy sources such as renewable energy can substantially reduce the carbon footprint of aerogel production. By harnessing solar, wind, or hydroelectric power, manufacturers can significantly decrease greenhouse gas emissions associated with aerogel manufacturing. Furthermore, adopting carbon capture and storage technologies can help offset remaining emissions, further enhancing the environmental sustainability of aerogel production. Overall, prioritizing sustainability in manufacturing practices and transitioning to cleaner energy sources are essential steps in mitigating the climate impact of aerogel production (Babcock-Jackson et al., 2023).

Resource Utilization: Aerogel production relies on raw materials like silica, carbon-based materials, polymers, and metallic nanoparticles, which can have significant environmental impacts during extraction and utilization. Implementing sustainable sourcing practices, such as sourcing raw materials from responsibly managed sources and promoting renewable alternatives, can reduce the environmental footprint of aerogel production. Additionally, material recycling and waste minimization strategies play a crucial role in mitigating environmental impact by reducing the demand for virgin resources and minimizing waste generation throughout the production process. By prioritizing resource efficiency, promoting circular economy principles, and adopting sustainable supply chain practices, aerogel manufacturers can minimize their environmental footprint and contribute to the overall sustainability of the materials industry (Ahmad et al., 2023).

Chemical Use and Waste Management: The synthesis of aerogel precursors often entails the use of various chemicals, solvents, and catalysts, which can contribute to environmental contamination and waste generation. However, adopting green chemistry principles can mitigate these risks by promoting the use of environmentally benign reagents and sustainable synthetic routes. Recycling solvents and implementing efficient waste management practices further reduce chemical use and minimize environmental pollution associated with aerogel production. By prioritizing the reduction of hazardous substances, minimizing waste generation, and promoting the use of renewable resources, green chemistry approaches offer a promising framework for enhancing the environmental sustainability of aerogel synthesis. These practices not only reduce the environmental footprint of aerogel production but also contribute to the overall advancement of sustainable manufacturing processes in the chemical industry (Mathew et al., 2024a; Zhang et al., 2023).

Figure 2. Environmental and health impact of aerogel

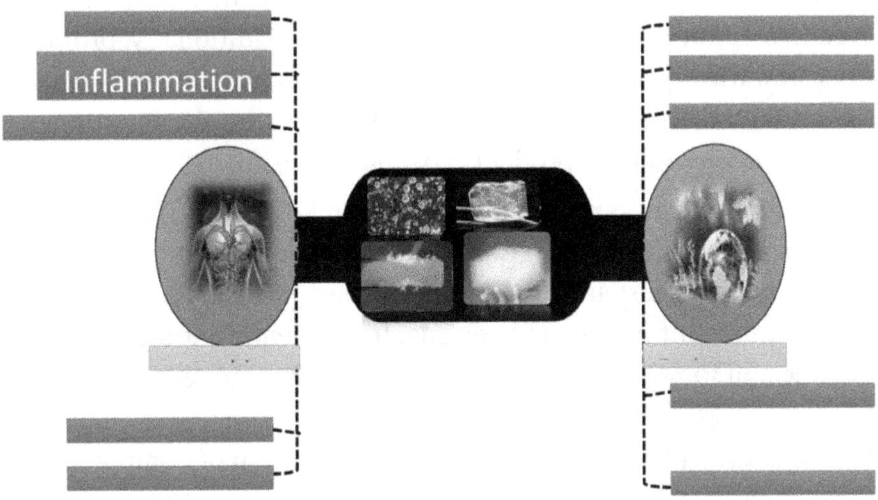

12.3.2 Main Environmental Considerations

The disposal of aerogel-based products at the end of their lifespan presents challenges due to their unique properties and composition. Traditional waste management practices may not be suitable for aerogels, leading to concerns regarding their environmental fate and potential long-term impacts (Bakhori et al., 2023).

Waste Generation: Aerogel-based products may pose challenges for disposal at the end of their lifespan due to their lightweight and porous nature. Traditional waste management methods are often ill-suited for aerogels, leading to potential accumulation in landfills or incineration. However, these disposal methods can result in environmental pollution and resource depletion. Aerogels' unique properties make them difficult to handle and recycle, further complicating the waste management process. As a result, there is a need for innovative disposal solutions that address the challenges posed by aerogel-based products. Implementing strategies such as advanced recycling technologies, alternative waste treatment methods, and sustainable end-of-life management practices can help minimize the environmental impact of aerogel waste disposal. Additionally, promoting product stewardship and consumer awareness can encourage responsible consumption and disposal practices, ultimately reducing the environmental footprint of aerogel-based products (Ahmad et al., 2023).

Environmental Persistence: Aerogels are known for their durability and resistance to degradation, posing a challenge for their management at the end of their lifespan. If not properly managed, aerogels can persist in the environment for extended periods. Implementing strategies such as product design for disassembly, material recovery, and reuse can significantly mitigate environmental impact. Designing aerogel-based products with disassembly in mind allows for easier separation of components, facilitating material recovery and recycling. Reusing aerogel-based products in other applications prolongs their lifespan, reducing the demand for new materials and minimizing waste generation. Additionally, incorporating circular economy principles, such as closed-loop systems and product stewardship, encourages responsible consumption and ensures the efficient use of resources. By adopting these strategies, the environmental persistence of aerogels can be mitigated, leading to a more sustainable approach to their utilization and disposal (Inobeme et al., 2022; Mahmoud et al., 2022).

Toxicity and Ecological Effects: While aerogels are generally considered inert and non-toxic, certain additives or impurities used in their production may pose environmental risks if released into the environment.

Assessing the potential toxicity of aerogel-based products is crucial to ensure environmental safety. This includes evaluating the composition of aerogels and identifying any hazardous substances present. Implementing appropriate disposal practices is essential to mitigate ecological effects. Methods such as controlled landfilling, incineration in specialized facilities, or recycling can help minimize the release of harmful substances into the environment. Additionally, promoting responsible use and handling of aerogel-based products can further reduce the risk of environmental contamination. Overall, proactive measures to assess toxicity and implement proper disposal practices are vital for minimizing the ecological impact of aerogel application (Keller et al., 2021; Inobeme et al., 2024).

Generally, the environmental impact assessment of aerogel application highlights the importance of considering the entire lifecycle of aerogel-based products, from production to disposal. Addressing key environmental considerations such as energy consumption, greenhouse gas emissions, resource utilization, chemical use, waste generation, and toxicity is essential for promoting sustainable aerogel utilization. Implementing measures to improve energy efficiency, reduce emissions, optimize resource utilization, minimize chemical use, and enhance end-of-life management practices can mitigate the environmental impact of aerogel application and promote a more sustainable future (Abdul-Khalil et al., 2023).

12.4 IDENTIFYING POTENTIAL EXPOSURE PATHWAYS ON AEROGEL APPLICATION

Potential exposure pathways to aerogel particles during application include inhalation, dermal contact, ingestion, and eye contact (**Table 2**). Inhalation can occur during aerogel handling or use, leading to respiratory irritation and lung inflammation. Dermal contact may result from direct exposure to aerogel surfaces, causing skin irritation and dryness. Accidental ingestion of aerogel particles poses a risk of gastrointestinal irritation. Eye contact with aerogel particles or dust may cause eye irritation and abrasion. Mitigation strategies include providing adequate ventilation and using personal protective equipment such as respirators to minimize inhalation exposure. For dermal contact, wearing gloves and protective clothing, and washing hands thoroughly after contact are recommended. Storing aerogel-based products securely, keeping them out of reach of children and pets, and avoiding eating or drinking in areas where aerogels are present can prevent ingestion. Using safety goggles or protective eyewear and flushing eyes with water if contact occurs can mitigate eye exposure (Boccia et al., 2023).

Table 2. Identifying potential exposure pathways on aerogel application

Exposure Pathway	Description	Potential Hazards	Mitigation Strategies	References
Inhalation	Inhalation of aerogel particles during handling or use	Respiratory irritation, lung inflammation	Provide adequate ventilation, use personal protective equipment (PPE) such as respirators, work in well-ventilated areas, minimize aerosolization	Chughtai & Khan, 2020
Dermal Contact	Direct contact with aerogel surfaces	Skin irritation, dryness	Wear appropriate PPE such as gloves and protective clothing, wash hands and skin thoroughly after contact, avoid prolonged skin exposure	Silva et al., 2022
Ingestion	Accidental ingestion of aerogel particles	Gastrointestinal irritation	Store aerogel-based products securely, keep out of reach of children and pets, avoid eating or drinking in areas where aerogels are present	Bakhori et al., 2023
Eye Contact	Contact with aerogel particles or dust in the eyes	Eye irritation, abrasion	Wear safety goggles or protective eyewear, flush eyes with water if contact occurs, seek medical attention if irritation persists	Rauchman et al., 2023

12.5 EVALUATING TOXICOLOGICAL EFFECTS ON AEROGEL APPLICATION

Evaluating the toxicological effects of aerogel application is crucial for ensuring the safety of workers and consumers. While aerogels are generally considered inert and non-toxic, potential risks may arise from exposure to aerogel particles or additives used during production. Toxicological studies aim to assess the potential health hazards associated with aerogel exposure through various routes, including inhalation, dermal contact, ingestion, and eye contact (Ferreira-Gonçalves et al., 2021; Mathew et al., 2024b).

Inhalation Exposure: Inhalation of aerogel particles is a primary concern for workers involved in aerogel production, handling, and application. Studies have shown that inhalation of silica nanoparticles, which are a common component of silica aerogels, can lead to respiratory irritation, lung inflammation, and fibrosis. Additionally, carbon-based aerogels may release carbon nanoparticles, posing similar respiratory risks. Toxicological evaluations involve exposing animal models or cell cultures to aerogel particles and assessing pulmonary effects, such as inflammatory responses, cytotoxicity, and lung function impairment (Bakhori et al., 2023).

Dermal Contact: Dermal exposure to aerogel surfaces may occur during handling or application, particularly in industries such as construction and manufacturing. While aerogels themselves are generally considered non-irritating to the skin, additives or impurities used in aerogel production may cause skin irritation or sensitization. Toxicological studies assess the potential dermal effects of aerogel exposure through patch testing on human volunteers or using animal models. Additionally, in vitro studies may evaluate the cytotoxicity of aerogel extracts on skin cells to determine potential adverse effects (Abdul-Khalil et al., 2020; Mathew et al. 2023a).

Ingestion: Accidental ingestion of aerogel particles is unlikely but may occur in occupational settings or through inadvertent ingestion of contaminated food or beverages. While aerogels are generally considered safe if ingested in small quantities, large doses may lead to gastrointestinal irritation or obstruction. Toxicological assessments involve administering aerogel particles to animal models via oral gavage and monitoring for signs of toxicity, such as changes in behavior, gastrointestinal function, and histopathological alterations in the gastrointestinal tract (Burdon et al., 2023).

Eye Contact: Eye exposure to aerogel particles or dust presents a potential risk of irritation, abrasion, or corneal damage. Toxicological evaluations play a crucial role in assessing the ocular irritancy potential of aerogels and determining the severity of eye irritation following exposure. Standardized tests such as the Draize test, which involves applying aerogel samples to the eyes of rabbits and observing for signs of irritation, are commonly used to evaluate acute eye irritation. Additionally, in vitro assays using human corneal epithelial cells provide valuable insights into the cytotoxicity and inflammatory responses induced by aerogel exposure. These studies aim to assess the potential for long-term damage to the eyes and inform risk management strategies to prevent ocular injuries. Mitigation measures for reducing the risk of eye exposure to aerogel particles include wearing protective eyewear, implementing engineering controls to minimize dust generation, and ensuring proper ventilation in areas where aerogel handling or application occurs (Rauchman et al., 2023).

Risk Assessment and Mitigation: Toxicological assessments play a critical role in informing risk management strategies and implementing appropriate mitigation measures to protect workers and consumers. Based on the findings of toxicological studies, exposure limits and protective measures, such as engineering controls, personal protective equipment (PPE), and safe handling procedures, can be established to minimize the risk of aerogel-related health effects. Additionally, regulatory agencies may require toxicity data as part of product registration or approval processes to ensure compliance with health and safety standards (Rout, and Sikdar, 2017; Abimbola et al., 2023).

Furthermore, evaluating the toxicological effects of aerogel application is essential for assessing potential health hazards and implementing appropriate risk management strategies. Toxicological studies assess the risks associated with inhalation, dermal contact, ingestion, and eye exposure to aerogel particles or additives. By identifying potential hazards and implementing mitigation measures, such as exposure controls and protective equipment, the safety of workers and consumers can be ensured during aerogel handling and application (Boccia et al., 2023).

12.6 IMPLICATIONS FOR SUSTAINABLE AEROGEL APPLICATION

As the utilization of aerogels continues to expand across various industries, it becomes imperative to address the environmental, social, and economic implications associated with their production, use, and disposal. Sustainable aerogel application entails integrating environmental stewardship, social responsibility, and economic viability throughout the entire lifecycle of aerogel-based products. This table (**Table 3**) discusses the key implications and strategies for promoting sustainable aerogel application. In general, sustainable aerogel application requires a holistic approach that integrates environmental stewardship, social responsibility, economic viability, and research and innovation. By addressing the environmental, social, and economic implications associated with aerogel production, use, and disposal, stakeholders can promote sustainable development and contribute to a more resilient and equitable future. Collaboration among industry, academia, government, and civil society is essential for driving progress towards sustainable aerogel application and realizing its full potential in addressing global challenges (Batisha, 2022; Mathew et al. 2024c).

Table 3. Implications for sustainable aerogel application

Aspect	Description	Strategies	Examples	References
Environmental	Minimizing environmental footprint throughout lifecycle	- Energy efficiency - Renewable energy - Resource efficiency	-Optimizing production processes to reduce energy consumption -Transitioning to solar or wind power for aerogel production - Implementing material recycling and waste minimization	Ragab et al., 2023
Social	Addressing social implications and ensuring equity	- Occupational health and safety - Community engagement - Equitable access	-Providing workers with adequate protective equipment - Engaging with local communities and stakeholders - Ensuring affordable access to aerogel-based technologies	Flynn et al., 202; Mathew et al. 2023b
Economic	Balancing environmental and social considerations with cost-effectiveness	- Cost optimization - Market development - Value chain integration	-Identifying cost-effective production methods - Investing in research to expand market opportunities - Collaborating with stakeholders to streamline production	Li & Liu, 2023
Research & Innovation	Advancing research and innovation to drive sustainability	- Material development - Process optimization - Life cycle assessment	-Developing aerogels with enhanced properties - Optimizing production processes for efficiency - Conducting comprehensive assessments to evaluate impacts	Liang et al., 2023

12.7 CONCLUSION

The environmental and health impact assessment of aerogel application highlights both opportunities and challenges associated with its widespread utilization. While aerogels offer significant benefits such as enhanced energy efficiency, lightweighting, and thermal insulation across various industries, their production, use, and disposal can pose environmental and health risks. Strategies to mitigate these impacts include implementing energy-efficient production processes, utilizing renewable energy sources, adopting sustainable sourcing practices, and enhancing end-of-life management practices. Additionally, addressing occupational hazards through proper training, equipment, and workplace controls is essential for ensuring the safety of workers involved in aerogel handling and application. Furthermore, promoting transparency, community engagement, and responsible consumption can foster trust and acceptance of aerogel technologies. Moving forward, continued research and innovation are critical for advancing sustainable aerogel application and maximizing its potential benefits while minimizing environmental and health risks. By incorporating environmental stewardship, social responsibility, economic viability, and research and innovation, stakeholders can contribute to a more sustainable and resilient future with aerogels playing a significant role in addressing global challenges.

REFERENCES

Abdul-Khalil, H. P. S., Jha, K., Yahya, E. B., Panchal, S., Patel, N., Garai, A., Kumari, S., & Jameel, M. (2023). Insights into the Potential of Biopolymeric Aerogels as an Advanced Soil-Fertilizer Delivery Systems. *Gels (Basel, Switzerland)*, *9*(8), 666. DOI: 10.3390/gels9080666 PMID: 37623121

Abimbola, O. F., Adetunji, C. O., Dare, B. J., Mathew, J. T., Inobeme, A., Ghazanfar, S., Olaniyan, O. T, Ajiboye, M. D., & Dauda, W. (2023). Packaging of Animal-Based Food Products: The Need for Intelligent Freshness Sensors. In book: Sensing and Artificial Intelligence Solutions for Food Manufacturing. CRC Press, 89-102. . DOI: 10.3390/gels9080666 PMID: 37623121

Ahmad, A., & Kamaruddin, M. A., H P S, A. K., Yahya, E. B., Muhammad, S., Rizal, S., Ahmad, M. I., Surya, I., & Abdullah, C. K. (. (2023). Recent Advances in Nanocellulose Aerogels for Efficient Heavy Metal and Dye Removal. *Gels (Basel, Switzerland)*, *9*(5), 416. DOI: 10.3390/gels9050416 PMID: 37233007

Ankley, G. T., Cureton, P., Hoke, R. A., Houde, M., Kumar, A., Kurias, J., Lanno, R., McCarthy, C., Newsted, J., Salice, C. J., Sample, B. E., Sepúlveda, M. S., Steevens, J., & Valsecchi, S. (2021). Assessing the Ecological Risks of Per- and Polyfluoroalkyl Substances: Current State-of-the Science and a Proposed Path Forward. *Environmental Toxicology and Chemistry*, *40*(3), 564–605. DOI: 10.1002/etc.4869 PMID: 32897586

Azeh, Y., Musah, M., Mathew, J. T., Badeggi, U. M., Estuyankpa, M. B., Muhammad, Z. D., Oluwaseyi, D. S., Umar, M. T., Abdulkarim, M. A., Muhammad, A. I., Ibrahim, A. A., Ismail, H., Muhammad, K. T., Umar, G., & Ahmed, T. (2023). In Situ Synthesis of PolyanilineNanohybrid and Formulation of Polyaniline/Carboxymethyl cellulose/Ethylene Glycol Nanocomposite: Study of its Conducting and Antibacterial Properties. *Journal of Materials and Polymer Science*, *3*(4), 1–9. DOI: 10.47485/2832-9384.1041

Babcock-Jackson, L., Konovalova, T., Krogman, J. P., Bird, R., & Díaz, L. L. (2023). Sustainable Fertilizers: Publication Landscape on Wastes as Nutrient Sources, Wastewater Treatment Processes for Nutrient Recovery, Biorefineries, and Green Ammonia Synthesis. *Journal of Agricultural and Food Chemistry, 71*(22), 8265–8296. DOI: 10.1021/acs.jafc.3c00454 PMID: 37219570

Bakhori, N. M., Ismail, Z., Hassan, M. Z., & Dolah, R. (2023). Emerging Trends in Nanotechnology: Aerogel-Based Materials for Biomedical Applications. *Nanomaterials (Basel, Switzerland), 13*(6), 1063. DOI: 10.3390/nano13061063 PMID: 36985957

Bakhori, N. M., Ismail, Z., Hassan, M. Z., & Dolah, R. (2023). Emerging Trends in Nanotechnology: Aerogel-Based Materials for Biomedical Applications. *Nanomaterials (Basel, Switzerland), 13*(6), 1063. DOI: 10.3390/nano13061063 PMID: 36985957

Batisha, A. (2022). Horizon scanning process to foresight emerging issues in Arabsphere's water vision. *Scientific Reports, 12*(1), 12709. DOI: 10.1038/s41598-022-16803-1 PMID: 35882917

Boccia, A. C., Pulvirenti, A., García-González, C. A., Grisi, F., & Neagu, M. (2023). Compendium of Safety Regulatory for Safe Applications of Aerogels. *Gels (Basel, Switzerland), 9*(11), 842. DOI: 10.3390/gels9110842 PMID: 37998932

Burdon, J., Budnik, L. T., Baur, X., Hageman, G., Howard, C. V., Roig, J., Coxon, L., Furlong, C. E., Gee, D., Loraine, T., Terry, A. V.Jr, Midavaine, J., Petersen, H., Bron, D., Soskolne, C. L., & Michaelis, S. (2023). Health consequences of exposure to aircraft contaminated air and fume events: A narrative review and medical protocol for the investigation of exposed aircrew and passengers. *Environmental Health : A Global Access Science Source, 22*(1), 43. DOI: 10.1186/s12940-023-00987-8 PMID: 37194087

Chughtai, A. A., & Khan, W. (2020). Use of personal protective equipment to protect against respiratory infections in Pakistan: A systematic review. *Journal of Infection and Public Health, 13*(3), 385–390. DOI: 10.1016/j.jiph.2020.02.032 PMID: 32146139

Ferreira-Gonçalves, T., Constantin, C., Neagu, M., Reis, C. P., Sabri, F., &Simón-Vázquez, R. (2021). Safety and efficacy assessment of aerogels for biomedical applications. *Biomedicine & pharmacotherapy = Biomedecine&pharmacotherapie, 144*, 112356. DOI: 10.1016/j.biopha.2021.112356

Flynn, M. A., Check, P., Steege, A. L., Sivén, J. M., & Syron, L. N. (2021). Health Equity and a Paradigm Shift in Occupational Safety and Health. *International Journal of Environmental Research and Public Health, 19*(1), 349. DOI: 10.3390/ijerph19010349 PMID: 35010608

Gatou, M. A., Vagena, I. A., Lagopati, N., Pippa, N., Gazouli, M., & Pavlatou, E. A. (2023). Functional MOF-Based Materials for Environmental and Biomedical Applications: A Critical Review. *Nanomaterials (Basel, Switzerland), 13*(15), 2224. DOI: 10.3390/nano13152224 PMID: 37570542

Ghica, M. E., Mandinga, J. G. S., Linhares, T., Almeida, C. M. R., & Durães, L. (2023). Improvement of the Mechanical Properties of Silica Aerogels for Thermal Insulation Applications through a Combination of Aramid Nanofibres and Microfibres. *Gels (Basel, Switzerland), 9*(7), 535. DOI: 10.3390/gels9070535 PMID: 37504414

Inobeme, A., Mathew, J.T., Adetunji, C.O., Agbugui, M.O., Osarenren, E.J., Oti, C., Ngonso, B.F., Inobeme, J., Oyedolapo, M.B., Bernard, E. &Jibrin, N.A. (2024). Biocomposites with graphene derivatives. In book: Advances in Biocomposites and their Applications, Elsevier Inc., 149-166. DOI: DOI: 10.1016/B978-0-443-19074-2.00005-8

Inobeme, A., Nayak, V., Mathew, J. T., Okonkwo, S., Ekwoba, L., Ajai, A. I., Bernard, E., Inobeme, J., Agbugui, M. M., & Singh, K. R. B. (2022). Chemometric approach in environmental pollution analysis: A critical review. *Journal of Environmental Management, 309*(114653), 1–18. DOI: 10.1016/j.jenvman.2022.114653 PMID: 35176568

Jin, R., Zhou, Z., Liu, J., Shi, B., Zhou, N., Wang, X., Jia, X., Guo, D., & Xu, B. (2023). Aerogels for Thermal Protection and Their Application in Aerospace. *Gels (Basel, Switzerland), 9*(8), 606. DOI: 10.3390/gels9080606 PMID: 37623061

Keller, J. G., Wiemann, M., Gröters, S., Werle, K., Vennemann, A., Landsiedel, R., & Wohlleben, W. (2021). Aerogels are not regulated as nanomaterials, but can be assessed by tiered testing and grouping strategies for nanomaterials. *Nanoscale Advances*, *3*(13), 3881–3893. DOI: 10.1039/D1NA00044F PMID: 36133012

Keller, J. G., Wiemann, M., Gröters, S., Werle, K., Vennemann, A., Landsiedel, R., & Wohlleben, W. (2021). Aerogels are not regulated as nanomaterials, but can be assessed by tiered testing and grouping strategies for nanomaterials. *Nanoscale Advances*, *3*(13), 3881–3893. DOI: 10.1039/D1NA00044F PMID: 36133012

Khan, N. R., Sharmin, T., & Bin Rashid, A. (2023). Exploring the Versatility of Aerogels: Broad Applications in Biomedical Engineering, Astronautics, Energy Storage, Biosensing, and Current Progress. *Heliyon*, *10*(1), e23102. DOI: 10.1016/j.heliyon.2023.e23102 PMID: 38163169

Koehle, A.P., Brumwell, S.L. &Seto, E.P. (2023). Microbial applications for sustainable space exploration beyond low Earth orbit. *npj Microgravity* 9, 47. DOI: 10.1038/s41526-023-00285-0

Lee, K. H., Arshad, Z., Dahshan, A., Alshareef, M., Alsulami, Q. A., Bibi, A., Lee, E.-J., Nawaz, M., Zubair, U., & Javid, A. (2023). Porous Aerogel Structures as Promising Materials for Photocatalysis, Thermal Insulation Textiles, and Technical Applications: A Review. *Catalysts*, *13*(9), 1286. DOI: 10.3390/catal13091286

Li, W., & Liu, Z. (2023). Social, Environmental, and Governance Factors on Supply-Chain Performance with Mediating Technology Adoption. *Sustainability (Basel)*, *15*(14), 10865. DOI: 10.3390/su151410865

Liang, Y., Li, C., Liu, Z., Wang, X., Zeng, F., Yuan, X., & Pan, Y. (2023). Decarbonization potentials of the embodied energy use and operational process in buildings: A review from the life-cycle perspective. *Heliyon*, *9*(10), e20190. DOI: 10.1016/j.heliyon.2023.e20190 PMID: 37810847

Mahmoud, L. A. M., Dos Reis, R. A., Chen, X., Ting, V. P., & Nayak, S. (2022). Metal-Organic Frameworks as Potential Agents for Extraction and Delivery of Pesticides and Agrochemicals. *ACS Omega*, *7*(50), 45910–45934. DOI: 10.1021/acsomega.2c05978 PMID: 36570238

Mathew, J. T., Adetunji, C. O., Inobeme, A., Musah, M., Azeh, Y., Otori, A. A., Shaba, E. Y., Mamman, A., & Tanko, E. (2023)b. Removal of Heavy Metals Using Bio-remedial Techniques. Springer Nature Switzerland AG 2023 M. P. Shah (ed.), Modern Approaches in Waste Bioremediation, 117-130. DOI: 10.1007/978-3-031-24086-7_6

Mathew, J. T., Inobeme, A., Azeh, Y., Musah, M., Abdulkadir, A., Shaba, E. Y., Etsuyankpa, M. B., Musa, T. S., Muhammad, A. I., Ismail, H., Kanwa, A. M., Mamman, A., & Hussaini, J. (2024). b. Carbon Capture and Storage Via Electrochemical and Bioelectrochemical Techniques: A Review. *Caliphate Journal of Science & Technology*, *2*(2), 141–158. DOI: 10.4314/cajost.v6i2.3

Mathew, J. T., Inobeme, A., Musah, M., Azeh, Y., Abdullahi, A., Shaba, E. Y., Salihu, A. M., Muhammad, E. B., Josiah, J. G., Jibrin, N. A., Ismail, H., Muhammad, A. I., Maurice, J., Mamman, A., & Ndamitso, M. M. (2024). a. A Critical Review of Green Approach on Wastewater Treatment Strategies. *Journal of Applied Science & Environmental Management*, *28*(2), 363–391. DOI: 10.4314/jasem.v28i2.9

Mathew, J. T., Musah, M., Azeh, Y., & Muhammed, M. (2023). a. Kinetic Study of Heavy Metals Removal from Pharmaceutical Wastewater Using Geopolymer/Fe_3O_4Nanocomposite. *Bima Journal of Science and Technology*, *7*(4), 152–163. DOI: 10.56892/bima.v7i4.539

Mathew, J. T., Musah, M., Azeh, Y., & Muhammed, M. (2024). c. Development of Fe3O4 Nanoparticles for the Removal of Some Toxic Metals from Pharmaceutical Wastewater. [CaJoST]. *Caliphate Journal of Science & Technology*, *6*(1), 26–34. DOI: 10.4314/cajost.v6i1.4

Nan, Z., Wei, W., Lin, Z., Chang, J., & Hao, Y. (2023). Flexible Nanocomposite Conductors for Electromagnetic Interference Shielding. *Nano-Micro Letters*, *15*(1), 172. DOI: 10.1007/s40820-023-01122-5 PMID: 37420119

Noman, M. T., Amor, N., Ali, A., Petrik, S., Coufal, R., Adach, K., & Fijalkowski, M. (2021). Aerogels for Biomedical, Energy and Sensing Applications. *Gels (Basel, Switzerland)*, *7*(4), 264. DOI: 10.3390/gels7040264 PMID: 34940324

Paraskevopoulou, P., Chriti, D., Raptopoulos, G., & Anyfantis, G. C. (2019). Synthetic Polymer Aerogels in Particulate Form. *Materials (Basel), 12*(9), 1543. DOI: 10.3390/ma12091543 PMID: 31083421

Ragab, A., Zouli, N., Abutaleb, A., Maafa, I. M., Ahmed, M. M., & Yousef, A. (2023). Environmental and Economic Benefits of Using Pomegranate Peel Waste for Insulation Bricks. *Materials (Basel), 16*(15), 5372. DOI: 10.3390/ma16155372 PMID: 37570075

Rauchman, S. H., Locke, B., Albert, J., De Leon, J., Peltier, M. R., & Reiss, A. B. (2023). Toxic External Exposure Leading to Ocular Surface Injury. *Vision (Basel), 7*(2), 32. DOI: 10.3390/vision7020032 PMID: 37092465

Rauchman, S. H., Locke, B., Albert, J., De Leon, J., Peltier, M. R., & Reiss, A. B. (2023). Toxic External Exposure Leading to Ocular Surface Injury. *Vision (Basel), 7*(2), 32. DOI: 10.3390/vision7020032 PMID: 37092465

Rout, B. K., & Sikdar, B. K. (2017). Hazard Identification, Risk Assessment, and Control Measures as an Effective Tool of Occupational Health Assessment of Hazardous Process in an Iron Ore Pelletizing Industry. *Indian Journal of Occupational and Environmental Medicine, 21*(2), 56–76. DOI: 10.4103/ijoem.IJOEM_19_16 PMID: 29540967

Silva, L. F. M. D., Almeida, A. G. A., Pascoal, L. M., Santos Neto, M., Lima, F. E. T., & Santos, F. S. (2022). Skin injuries due to Personal Protective Equipment and preventive measures in the COVID-19 context: an integrative review. Lesões de peleporEquipamentos de Proteção Individual e medidaspreventivas no contexto da COVID-19: revisãointegrativa. *Revista Latino-Americana de Enfermagem, 30*, e3551. DOI: 10.1590/1518-8345.5636.3551 PMID: 35476010

Sozcu, S., Venkataraman, M., Wiener, J., Tomkova, B., Militky, J., & Mahmood, A. (2023). Incorporation of Cellulose-Based Aerogels into Textile Structures. *Materials (Basel), 17*(1), 27. DOI: 10.3390/ma17010027 PMID: 38203881

Vrabič-Brodnjak, U. (2024). Hybrid Materials of Bio-Based Aerogels for Sustainable Packaging Solutions. *Gels (Basel, Switzerland), 10*(1), 27. DOI: 10.3390/gels10010027 PMID: 38247750

Wang, Z., Liu, L., Zhang, Y., Huang, Y., Liu, J., Zhang, X., Liu, X., Teng, H., Zhang, X., Zhang, J., & Yang, H. (2023). A Review of Graphene-Based Materials/Polymer Composite Aerogels. *Polymers*, *15*(8), 1888. DOI: 10.3390/polym15081888 PMID: 37112034

Xue, T., Zhu, C., Yu, D., Zhang, X., Lai, F., Zhang, L., Zhang, C., Fan, W., & Liu, T. (2023). Fast and scalable production of crosslinked polyimide aerogel fibers for ultrathin thermoregulating clothes. *Nature Communications*, *14*(1), 8378. DOI: 10.1038/s41467-023-43663-8 PMID: 38104160

Zhang, Y., Poon, K., Masonsong, G. S. P., Ramaswamy, Y., & Singh, G. (2023). Sustainable Nanomaterials for Biomedical Applications. *Pharmaceutics*, *15*(3), 922. DOI: 10.3390/pharmaceutics15030922 PMID: 36986783

Chapter 8
Exploring Future Perspectives in Aerogel-Based Research

John Tsado Mathew
https://orcid.org/0000-0001-9514-5330
Department of Chemistry, Ibrahimm Badamasi Babangida Lapai, Nigeria

Monday Musah
https://orcid.org/0000-0002-3224-1664
Department of Chemistry, Ibrahim Badamasi Babangida University, Lapai, Nigeria

Yakubu Azeh
https://orcid.org/0000-0002-3818-674X
Department of Chemistry, Ibrahim Badamasi Babangida University, Lapai, Nigeria

Elijah Yanda Shaba
Department of Chemistry, Federal University of Technology, Minna, Nigeria

Etsuyankpa Bini Muhammad
Department of Chemistry, Fedral University, Lafia, Nigeria

Jonathan Hussaini
https://orcid.org/0009-0004-9216-8691
Department of Chemistry, Ibrahim Badamasi Babangida University, Lapai, Nigeria

Musa Safiyanu Tanko
Department of Chemistry, Federal University, Lafia, Nigeria

Amos Mamman
Department of Chemistry, Ibrahim Badamasi Babangida University, Lapai, Nigeria

DOI: 10.4018/979-8-3693-5971-6.ch008

ABSTRACT

Aerogels, with their exceptional properties including ultra-low density, high surface area, and superior thermal insulation, have garnered significant attention across various fields of research and application. In this abstract, we delve into the future perspectives of aerogel-based research, highlighting emerging trends and potential directions. One promising avenue is the advancement of aerogel manufacturing techniques, aiming for scalability, cost-effectiveness, and environmental sustainability. Additionally, the integration of aerogels into novel applications such as energy storage, catalysis, and biomedical devices presents exciting opportunities for innovation. Furthermore, exploring the potential of aerogels in emerging fields like aerospace, wearable technology, and environmental remediation holds promise for transformative breakthroughs. This abstract provides a glimpse into the future landscape of aerogel research, emphasizing the importance of continued exploration and innovation in harnessing the remarkable properties of these remarkable materials

13.1 INTRODUCTION

Aerogels, colloquially referred to as "frozen smoke" or "solid air," have captivated scientists and engineers for decades due to their remarkable properties and potential applications across various fields. These ultra-lightweight materials, composed of a porous network of interconnected nanoparticles suspended in a gas, boast unparalleled characteristics such as extremely low density, high porosity, large surface area, and exceptional thermal insulation. Since their discovery in the 1930s by Samuel Kistler, aerogels have undergone significant advancements, leading to a surge in research activities aimed at unlocking their full potential (Lee et al., 2023).

The inception of aerogel research can be traced back to Kistler's pioneering work on the synthesis of silica aerogels through the supercritical drying of gels. Initially developed for fundamental studies in colloid chemistry, aerogels soon garnered attention for their unique properties, including their ultra-low density (typically less than 0.1 g/cm^3) and high surface area (ranging from several hundred to over a thousand square meters per gram). These properties arise from the nanoscale structure of aerogels, where the interconnected nanoparticles form a highly porous framework with pore sizes on the order of nanometers. Over the years, researchers have expanded the repertoire of aerogel materials beyond silica to include a wide range of compositions such as metal oxides, polymers, carbon-based materials, and composites. This diversification has broadened the scope of aerogel applications and spurred interest in tailoring their properties to meet specific requirements in various industries (Payanda et al., 2023).

The significance of aerogel-based research lies in the transformative potential of these materials across numerous fields, ranging from energy and environmental sustainability to aerospace and biomedical applications. One of the most notable attributes of aerogels is their exceptional thermal insulation properties, which stem from the combination of low density and high porosity. This makes aerogels ideal candidates for insulation in buildings, refrigerators, and cryogenic systems, where minimizing heat transfer is essential for energy efficiency (Bakhori et al., 2023).

Furthermore, the large surface area and tunable pore structure of aerogels offer opportunities for applications in gas storage, catalysis, and filtration. In the realm of energy storage, aerogels hold promise as electrode materials for supercapacitors, batteries, and fuel cells, owing to their high surface area and excellent electrical conductivity. In addition to their utility in traditional engineering applications, aerogels have found niche roles in cutting-edge technologies such as aerospace and wearable devices. The lightweight nature of aerogels makes them attractive for use in aerospace structures, where weight reduction is critical for improving fuel efficiency and payload capacity. Similarly, in the realm of wearable technology, aerogels can enhance comfort and performance by providing lightweight yet effective insulation in clothing and protective gear (Yang et al., 2023).

Moreover, the biocompatibility and porosity of certain aerogel formulations make them suitable for biomedical applications such as drug delivery, tissue engineering, and wound healing. By precisely controlling the composition and pore structure, researchers can tailor aerogels to mimic the extracellular matrix of tissues or serve as carriers for therapeutic agents, opening new avenues for regenerative medicine and personalized healthcare. In light of these diverse applications and the ever-growing demand for sustainable materials, aerogel-based research is poised to play a crucial role in addressing pressing societal challenges such as climate change, energy security, and healthcare. However, despite significant progress in aerogel synthesis, characterization, and application development, several challenges remain to be addressed to fully exploit the potential of these extraordinary materials (Nita et al., 2020).

In the subsequent sections of this paper, we will explore the future perspectives in aerogel-based research, focusing on advancements in manufacturing techniques, emerging applications, interdisciplinary collaboration, and the broader implications for science and technology. Through a comprehensive analysis of current trends and future directions, we aim to provide insights into the transformative role of aerogels in shaping the future of materials science and engineering (García-González et al., 2019).

13.2 ADVANCEMENTS IN AEROGEL MANUFACTURING TECHNIQUES

Advancements in aerogel manufacturing techniques have been pivotal in enhancing the scalability, efficiency, and versatility of aerogel production for various applications (**Table 1**). The sol-gel method, a widely employed technique, involves the controlled chemical synthesis of a sol followed by gelation to form the aerogel structure. This approach offers precise control over composition and structure, enabling the creation of aerogels with tailored properties suitable for insulation, catalysis, and biomedical devices.

Supercritical drying, another significant technique, involves the removal of solvent from the gel using supercritical fluid, preserving the porous structure of the aerogel. This method ensures uniform drying and is particularly advantageous for applications requiring thermal insulation and energy storage. Freeze drying, characterized by freezing the gel followed by the sublimation of the solvent, maintains the nanoporous structure of aerogels and is scalable for large-scale production. This technique finds applications in aerospace materials and drug delivery systems where maintaining the integrity of the aerogel structure is critical. Template-assisted synthesis employs templates to control pore structure, resulting in aerogels with tailored porosity and high surface area. This technique has been instrumental in applications such as gas storage, filtration, and tissue engineering, where specific pore structures are desired for optimal performance (Guastaferro et al., 2021).

Table 1. Outlining advancements in aerogel manufacturing techniques

Manufacturing Technique	Description	Advantages	Applications	References
Sol-Gel Method	Chemical synthesis involving sol formation and gelation	Precise control over composition and structure	Insulation, catalysis, biomedical devices	Catauro & Ciprioti, 2021
Supercritical Drying	Removal of solvent from gel using supercritical fluid	Preserves porous structure, uniform drying	Thermal insulation, energy storage	Payanda et al., 2023
Freeze Drying	Freezing the gel followed by sublimation of solvent	Maintains nanoporous structure, scalable	Aerospace materials, drug delivery systems	Guastaferro et al., 2021
Template-Assisted Synthesis	Using templates to control pore structure	Tailored porosity, high surface area	Gas storage, filtration, tissue engineering	Li et al., 2022

13.3 EMERGING APPLICATIONS OF AEROGELS

Aerogels, with their unique combination of properties including ultra-low density, high surface area, and excellent thermal insulation, are finding diverse and innovative applications across various industries. While aerogels have been traditionally used in insulation and filtration, emerging applications are leveraging their exceptional properties to address contemporary challenges and drive technological advancements (Noman et al., 2021).

Energy Storage: Aerogels hold promise as advanced materials for energy storage devices such as supercapacitors, batteries, and fuel cells. Due to their high surface area and porosity, aerogels can serve as efficient electrodes, providing a large interface for electrochemical reactions. Additionally, their lightweight nature minimizes the overall weight of energy storage systems, making them suitable for portable and aerospace applications. Research efforts are focused on optimizing the electrochemical performance of aerogels, including enhancing conductivity, stability, and charge storage capacity, to unlock their full potential in next-generation energy storage technologies (Khan et al., 2023).

Catalysis: The unique properties of aerogels make them attractive candidates for catalytic applications. Their high surface area and porous structure facilitate efficient mass transport and provide abundant active sites for catalytic reactions. Aerogel-supported catalysts have shown promising results in various catalytic processes, including environmental remediation, chemical synthesis, and fuel conversion. By tuning the composition and structure of aerogel catalysts, researchers aim to enhance catalytic activity, selectivity, and durability, paving the way for sustainable and energy-efficient catalytic processes (Almeida et al., 2024).

Biomedical Devices: Aerogels are gaining attention in the biomedical field for their biocompatibility, tunable porosity, and controlled drug release properties. Aerogel-based scaffolds are being investigated for tissue engineering applications, where they can provide a three-dimensional structure for cell growth and tissue regeneration. Additionally, aerogels are explored as carriers for drug delivery systems, enabling controlled release of therapeutic agents with tailored release kinetics. Moreover, aerogels exhibit potential in wound dressing materials, where their high porosity and moisture-absorbing capacity promote wound healing and prevent infection. Continued research in aerogel-based biomedical devices holds promise for addressing unmet medical needs and advancing regenerative medicine (Adetunji et al., 2021; Adetunji et al., 2023a).

Aerospace Materials: In the aerospace industry, where lightweight materials are crucial for reducing fuel consumption and enhancing aircraft performance, aerogels offer unique advantages. Aerogel-based composites are being developed for aerospace applications, including structural components, thermal insulation, and aerodynamic surfaces. The low density of aerogels contributes to weight savings without compromising mechanical strength, making them ideal candidates for aerospace materials. Additionally, aerogels exhibit excellent thermal insulation properties, enabling effective thermal management in spacecraft and aircraft systems.

As research progresses, aerogel-based aerospace materials are expected to play a significant role in improving efficiency, sustainability, and safety in the aerospace industry (Jin et al., 2023).

Environmental Remediation: Aerogels are being explored for environmental remediation applications, including water purification, air filtration, and hazardous waste remediation. The high surface area and porous structure of aerogels facilitate adsorption and removal of contaminants from water and air streams. Functionalized aerogels with specific affinity for target pollutants can selectively capture pollutants, leading to efficient remediation processes. Additionally, aerogels can serve as adsorbents for capturing greenhouse gases and volatile organic compounds, contributing to efforts to mitigate air pollution and climate change. By harnessing the unique properties of aerogels, environmental remediation technologies can achieve enhanced efficiency, cost-effectiveness, and sustainability (Mathew et al., 2023a; Mathew et al., 2023b).

Wearable Technology: Aerogels are increasingly being integrated into wearable technology devices to enhance comfort, performance, and functionality. Aerogel-based materials offer lightweight and flexible insulation solutions for clothing, protective gear, and wearable electronics. By incorporating aerogels into garment linings or protective layers, manufacturers can improve thermal insulation, moisture management, and impact resistance without adding bulk or weight. Moreover, aerogels can be tailored to specific requirements, such as breathability, flexibility, and durability, to meet the diverse needs of wearable technology applications. As the demand for wearable technology continues to grow, aerogels are poised to play a significant role in enhancing user experience and enabling innovative wearable devices (Dolez et al., 2022).

In general, aerogels are versatile materials with a wide range of emerging applications across industries such as energy storage, catalysis, biomedical devices, aerospace materials, environmental remediation, and wearable technology. Continued research and innovation in aerogel materials and manufacturing techniques are essential for unlocking their full potential and addressing global challenges in energy, healthcare, environmental sustainability, and beyond (Khan et al., 2023).

13.4 AEROGELS IN AEROSPACE AND WEARABLE TECHNOLOGY

Aerogels are increasingly being utilized in both aerospace and wearable technology applications, leveraging their unique properties to enhance performance and functionality.

In aerospace, where weight reduction is critical for fuel efficiency and payload capacity, aerogels offer an attractive solution. Aerogel-based materials can be incorporated into structural components, thermal insulation systems, and aerodynamic surfaces. Their ultra-low density enables significant weight savings without compromising mechanical strength, making them ideal for use in aircraft and spacecraft. Additionally, aerogels' exceptional thermal insulation properties contribute to effective thermal management in extreme environments encountered during flight. By utilizing aerogels in aerospace materials, manufacturers can improve efficiency, reduce operational costs, and enhance safety (Jin et al., 2023).

In wearable technology, aerogels are utilized to enhance comfort, insulation, and protection. Aerogel-based materials offer lightweight and flexible solutions for clothing, footwear, and protective gear. By incorporating aerogels into garment linings or padding, manufacturers can provide superior thermal insulation, moisture management, and impact resistance. Aerogel-based insulation enables wearers to stay warm in cold climates while remaining lightweight and mobile. Moreover, aerogels can be tailored to specific requirements, such as breathability, flexibility, and durability, to meet the diverse needs of wearable technology applications. Whether integrated into outerwear, gloves, or even smart fabrics, aerogels play a crucial role in improving wearer comfort and performance in various environments (Lee et al., 2023).

Overall, aerogels offer significant advantages in aerospace and wearable technology applications, where lightweight, durable, and high-performance materials are essential. Continued research and innovation in aerogel materials and manufacturing techniques will further expand their use in these industries, driving advancements in aerospace technology, enhancing user experience in wearable devices, and paving the way for future innovations (Fan et al., 2024).

13.4.1 Aerospace Applications

Aerogels are increasingly finding diverse applications in the aerospace industry, owing to their unique combination of properties such as ultra-low density, high porosity, and exceptional thermal insulation. These properties make aerogels attractive materials for various aerospace applications, ranging from structural components to thermal protection systems. One significant aerospace application of aerogels is in lightweight structural materials. Aerogel-based composites can be used to manufacture aircraft components such as wings, fuselage panels, and interior structures. By incorporating aerogels into composite materials, manufacturers can achieve significant weight savings without compromising structural integrity. This reduction in weight translates to improved fuel efficiency, increased payload capacity, and enhanced overall performance of aircraft. Additionally, aerogels are utilized in thermal protection systems for spacecraft and re-entry vehicles. During atmospheric re-entry, vehicles experience extreme temperatures that can reach thousands of degrees Celsius (Sheng et al., 2023). Aerogels' exceptional thermal insulation properties make them ideal candidates for thermal shielding, protecting spacecraft components and payloads from the intense heat generated during re-entry. Aerogel-based thermal protection systems enable safer and more reliable space missions by effectively managing thermal loads and minimizing heat transfer to sensitive equipment (Jin et al., 2023).

Moreover, aerogels are employed in aerospace applications for cryogenic insulation. In cryogenic environments encountered in space missions and rocket propulsion systems, conventional insulation materials may fail to provide adequate thermal protection due to their limited effectiveness at extremely low temperatures. Aerogels, with their low thermal conductivity and high porosity, offer superior insulation performance even at cryogenic temperatures, making them ideal candidates for cryogenic insulation applications. Aerogel-based insulation systems enable efficient storage and transport of cryogenic fluids such as liquid hydrogen and liquid oxygen, essential for space exploration and satellite propulsion. Furthermore, aerogels are investigated for use in aerospace materials with multifunctional properties. Researchers are exploring the integration of aerogels into advanced materials capable of providing structural support, thermal insulation, and electromagnetic shielding in a single component. These multifunctional aerospace materials offer numerous advantages, including reduced complexity, weight, and cost, while enhancing overall performance and reliability. Generally, aerogels play a crucial role in various aerospace applications, including lightweight structural materials, thermal protection systems, cryogenic insulation, and multifunctional materials. Their unique properties make them indispensable materials for improving aircraft and spacecraft performance, enhancing safety, and enabling advanced space missions. Continued research and development in aerogel materials and manufacturing techniques will further expand their applications in the aerospace industry, driving innovation and progress in aerospace technology (Sambucci et al., 2023).

13.4.2 Wearable Technology

Aerogels are revolutionizing the field of wearable technology, offering lightweight, flexible, and high-performance materials that enhance comfort, insulation, and protection for users. With their unique properties, aerogels are increasingly integrated into various wearable devices, ranging from clothing and footwear to smart fabrics and protective gear. One of the primary applications of aerogels in wearable technology is in thermal insulation. Aerogel-based materials provide superior thermal insulation while remaining lightweight and flexible, making them ideal for cold weather clothing, outdoor gear, and winter sports equipment. By incorporating aerogels into garment linings or insulation layers, manufacturers can keep wearers warm and comfortable in harsh environments without adding bulk or weight (Seesaard & Wongchoosuk, 2023).

Additionally, aerogels are utilized in footwear to enhance insulation and cushioning. Aerogel insoles offer lightweight yet effective insulation against cold surfaces, preventing heat loss and ensuring comfort in cold weather conditions. Furthermore, aerogel inserts can provide impact resistance and shock absorption, reducing the risk of foot fatigue and injuries during physical activities. Moreover, aerogels are integrated into smart fabrics and wearable electronics to improve functionality and performance. Aerogel-based materials can serve as lightweight and flexible substrates for embedding sensors, actuators, and electronic components in clothing, allowing for seamless integration of wearable technology into everyday attire. Aerogel-based fabrics also offer enhanced breathability and moisture management, ensuring wearer comfort and performance during prolonged use (Greszta et al., 2022; Olorunsola et al. 2024).

In addition to thermal insulation and smart textiles, aerogels are utilized in protective gear for sports, industrial, and military applications. Aerogel-based padding and armor provide lightweight yet effective protection against impact, abrasion, and puncture injuries. By incorporating aerogels into helmets, body armor, and protective clothing, manufacturers can improve wearer safety without sacrificing mobility or comfort. Overall, aerogels are driving innovation in wearable technology, offering versatile materials that enhance comfort, insulation, and protection for users. With their lightweight, flexible, and high-performance properties, aerogels are revolutionizing the design and functionality of wearable devices, enabling advancements in sports equipment, outdoor gear, smart textiles, and protective clothing. As research and development in aerogel materials continue to advance, the potential applications of aerogels in wearable technology are expected to expand further, leading to safer, more comfortable, and more functional wearable devices for diverse applications (Jin et al., 2023).

13.5 ENVIRONMENTAL REMEDIATION

Environmental remediation is a critical area where aerogels are increasingly being explored for their potential to address pollution and contamination challenges. With their unique properties such as high surface area, tunable pore structure, and excellent adsorption capabilities, aerogels offer versatile solutions for various environmental remediation applications, including water purification, air filtration, and hazardous waste management (Romero-Montero et al., 2023).

Water Purification: Aerogels hold promise for water purification applications due to their ability to efficiently adsorb contaminants and pollutants from water sources. Their high surface area and porous structure provide ample binding sites for target molecules, enabling effective removal of heavy metals, organic compounds, and microorganisms from contaminated water. Aerogel-based adsorbents can be tailored to specific contaminants through surface functionalization or modification, enhancing selectivity and efficiency in water treatment processes. Additionally, aerogels offer advantages such as rapid adsorption kinetics, regeneration capability, and resistance to fouling, making them suitable for continuous operation in water treatment systems. From industrial wastewater treatment to portable water purification devices for disaster relief efforts, aerogels have the potential to play a significant role in ensuring access to clean and safe water resources (Mathew et al., 2024a).

Air Filtration: Aerogels are also utilized for air filtration applications to remove particulate matter, volatile organic compounds (VOCs), and other airborne pollutants from indoor and outdoor environments. Their high porosity and interconnected pore structure enable efficient capture and retention of airborne particles and gases, leading to improved indoor air quality and reduced health risks associated with air pollution. Aerogel-based filters exhibit high filtration efficiency, low pressure drop, and long service life, making them suitable for various air purification systems, including HVAC systems, automotive cabin filters, and personal respirators. Moreover, aerogel filters can be integrated into sustainable building designs to enhance energy efficiency and indoor environmental quality, contributing to healthier and more sustainable living environments (Shao et al., 2022; Mathew et al. 2023c).

Hazardous Waste Management: Aerogels offer innovative solutions for hazardous waste management and remediation of contaminated sites. Their high adsorption capacity and chemical stability make them effective sorbents for capturing and immobilizing hazardous substances such as heavy metals, radioactive elements, and organic pollutants. Aerogel-based materials can be deployed in remediation technologies such as adsorption columns, permeable reactive barriers, and in-situ stabilization/solidification techniques to mitigate risks associated with contaminated soil, groundwater, and industrial waste streams. Furthermore, aerogels can be functionalized with specific chemical groups or nanoparticles to enhance their affinity for target contaminants and improve remediation efficiency. By facilitating the containment, immobilization, and remediation of hazardous wastes, aerogels contribute to environmental protection, ecosystem restoration, and sustainable development efforts (Franco et al., 2021; Adetuyi et al. 2024).

Furthermore, aerogels hold significant promise for environmental remediation applications, offering efficient, cost-effective, and sustainable solutions for addressing water and air pollution, as well as managing hazardous wastes. Continued research and development in aerogel materials and remediation technologies will further expand their use in environmental protection and restoration efforts, contributing to the preservation of natural resources, protection of human health, and mitigation of environmental risks. As global environmental challenges continue to escalate, the versatility and effectiveness of aerogels in remediation applications make them invaluable tools for sustainable environmental management and conservation (Saleh et al., 2023).

13.6 INTERDISCIPLINARY COLLABORATION IN AEROGEL RESEARCH

Aerogels, with their unique properties and diverse applications, are at the forefront of interdisciplinary research, bringing together scientists, engineers, and experts from various fields to address complex challenges and drive innovation (**Table 2**). Interdisciplinary collaboration in aerogel research is essential for advancing fundamental understanding (**figure 1**), developing novel materials and applications, and addressing societal needs across multiple domains (Chen et al., 2023).

Figure 1. Collaboration in aerogel research

Materials Science and Engineering: Materials scientists and engineers play a central role in aerogel research, focusing on the synthesis, characterization, and optimization of aerogel materials for specific applications. They explore novel aerogel formulations, fabrication techniques, and processing methods to tailor aerogel properties such as pore structure, surface chemistry, and mechanical strength. By understanding the relationship between composition, structure, and performance, materials scientists optimize aerogel materials for diverse applications ranging from insulation and catalysis to biomedical devices and environmental remediation (Bakhori et al., 2023).

Chemistry and Nanotechnology: Chemists and nanotechnologists contribute to aerogel research by elucidating the underlying chemical and physical principles governing aerogel formation and behavior at the nanoscale. They investigate the chemistry of precursor materials, sol-gel processes, and gelation mechanisms to control aerogel morphology, porosity, and surface properties. Additionally, they leverage nanotechnology techniques such as nanoparticle synthesis, surface modification, and self-assembly to engineer advanced aerogel materials with tailored functionalities and enhanced performance. Collaborations between chemists, nanotechnologists, and materials scientists drive innovation in aerogel synthesis, enabling the development of next-generation aerogel materials with unprecedented properties and applications (Inobeme et al., 2023).

Physics and Thermodynamics: Physicists and thermodynamicists contribute to aerogel research by studying the thermal, mechanical, and optical properties of aerogel materials and elucidating their underlying physical mechanisms. They investigate phenomena such as thermal conductivity, sound absorption, light scattering, and mechanical behavior to understand aerogel performance under different conditions. By employing theoretical models, computational simulations, and experimental techniques, physicists provide insights into aerogel behavior at the microstructural and macroscopic levels. Their contributions enhance our understanding of aerogel properties and guide the design of aerogel materials for specific applications in energy storage, aerospace, and thermal insulation (Ferreira-Gonçalves et al., 2022; Adetunji et al., 2023b).

Biomedical Sciences and Biotechnology: Biomedical scientists and biotechnologists explore the potential of aerogels for biomedical applications, including drug delivery, tissue engineering, and wound healing. They investigate the biocompatibility, bioactivity, and degradation kinetics of aerogel materials in biological environments. Additionally, they explore strategies to functionalize aerogel surfaces, incorporate bioactive molecules, and control release kinetics for targeted drug delivery and tissue regeneration. Collaborations between biomedical scientists, materials engineers, and clinicians enable the development of aerogel-based biomaterials with tailored properties and functionalities for applications in regenerative medicine, drug delivery, and medical devices (Inobeme et al., 2024).

Environmental Science and Sustainability: Environmental scientists and sustainability experts focus on harnessing the potential of aerogels for environmental remediation, pollution control, and sustainable technologies. They investigate the adsorption, absorption, and catalytic properties of aerogel materials for removing pollutants from air, water, and soil. Additionally, they explore the use of aerogels in energy-efficient technologies such as carbon capture, water purification, and sustainable agriculture.

Collaborations between environmental scientists, materials engineers, and policymakers drive innovation in aerogel-based solutions for addressing environmental challenges and advancing sustainability goals (Finny 2024).

Industry and Technology Transfer: Industry partners and technology transfer specialists facilitate the translation of aerogel research from the laboratory to commercial applications. They collaborate with academic researchers, government agencies, and funding organizations to identify market opportunities, assess technological feasibility, and navigate regulatory requirements. Additionally, they provide expertise in scaling up production, optimizing manufacturing processes, and integrating aerogel materials into existing products and systems. Collaborations between academia, industry, and government accelerate the commercialization of aerogel technologies, fostering innovation, economic growth, and societal impact (Sena-Torralba et al. 2022).

In addition, interdisciplinary collaboration is essential for advancing aerogel research and realizing the full potential of aerogel materials across diverse applications. By bringing together scientists, engineers, and experts from various disciplines, interdisciplinary collaborations drive innovation, address complex challenges, and accelerate the translation of aerogel research into practical solutions for societal needs. As aerogel research continues to evolve, interdisciplinary collaboration will remain critical for pushing the boundaries of scientific knowledge, developing novel materials and technologies, and creating positive impacts on society and the environment (Tawiah et al., 2023).

Table 2. Outlining interdisciplinary collaboration in aerogel research

Discipline	Contribution	Focus Area	Key Activities	Examples of Collaborations	References
Materials Science & Engineering	Synthesis, characterization, optimization of aerogels	Tailoring properties, fabrication techniques	Novel formulations, processing methods	Developing aerogels for thermal insulation	Zhan et al., 2023
Chemistry & Nanotechnology	Understanding chemical processes, nanoscale behavior	Precursor chemistry, surface modification	Nanoparticle synthesis, self-assembly	Engineering functionalized aerogel materials	Mathew et al., 2024b
Physics & Thermodynamics	Studying thermal, mechanical, optical properties	Thermal conductivity, light scattering	Theoretical modeling, computational simulations	Investigating aerogel behavior under conditions	Liu et al., 2022
Biomedical Sciences & Biotechnology	Exploring applications in drug delivery, tissue engineering	Biocompatibility, bioactivity, degradation kinetics	Functionalization strategies, release kinetics	Developing aerogel-based biomaterials	Thang et al., 2023
Environmental Science & Sustainability	Investigating applications in environmental remediation	Adsorption, absorption, catalytic properties	Pollution control, carbon capture	Utilizing aerogels for water purification	Li et al., 2022

13.7 CONCLUSION

The future of aerogel-based research is brimming with promise and potential across multiple

Figure 2. Future trend in aerogel based research

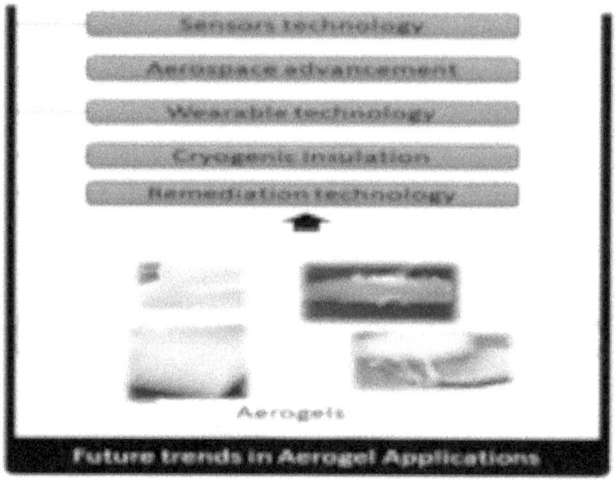

As interdisciplinary collaboration continues to thrive, researchers are poised to unlock new avenues of innovation and address pressing societal challenges. Advancements in aerogel manufacturing techniques, coupled with emerging applications in areas such as energy storage, catalysis, biomedical devices, aerospace, environmental remediation, and wearable technology, underscore the versatility and significance of aerogels in shaping the future of materials science and engineering. Moreover, ongoing efforts to optimize aerogel properties, enhance performance, and scale up production are paving the way for widespread adoption and commercialization of aerogel technologies. By harnessing the remarkable properties of aerogels and leveraging interdisciplinary approaches, researchers are driving transformative breakthroughs with far-reaching implications for sustainability, efficiency, and quality of life. As we embark on this exciting journey of exploration and innovation, collaboration, creativity, and cross-disciplinary exchange will remain essential for realizing the full potential of aerogels and unleashing their transformative impact on society, industry, and the environment. The future of aerogel-based research holds boundless possibilities, offering opportunities to address complex challenges and create a brighter, more sustainable future for generations to come.

REFERENCES

Adetunji, C. O., Egbuna, C., Ajiboye, A. A. A., Ajayi, O. O., Dauda, W. P., Ghazanfar, S., Ogundolie, F. A., Mathew, J. T., Adetunji, J. B., Inobeme, A., OlotuTitilayo, Olaniyan, O. T., Ijabadeniyi, O. A., Aborode, A. T., Ficai, A., Akram, M., Bodunrinde, R. E., Laila, U., Olaniyan, O. P., Micheal, O. S., Aliyu, A., Akinbo, O. &Adetuyi, B. O. (2023)a. Biogenic nanomaterials with diverse biological activities in the food and biomedical industries, Evaporative Coolers for the Postharvest Management of Fruits and Vegetables, Academic Press, 395-420, . DOI: 10.1007/978-981-99-8357-5_16

Adetunji, C. O., Mathew, J. T., Inobeme, A., Taiwo, O., Olaniyan, O.,Shakira,G., Wadazani,D.(2023)b.Application of Biosensors for Detection and Monitoring of Water Quality.In book: Organic and Inorganic Materials Based Sensors, John Wiley & Sons, Inc., 1065-1077, DOI: 10.1007/978-981-99-8357-5_16

Adetunji, C. O., Olaniyan, O. T., Inobeme, A., Mathew, J. T., & Ahamed, M. I. (2021). Waterborne Polyurethanes for Biomedical Applications. In: Inamuddin, Boddula R., Khan A. (eds) Sustainable Production and Applications of Waterborne Polyurethanes. Advances in Science, Technology & Innovation (IEREK Interdisciplinary Series for Sustainable Development). Springer, Cham. DOI: 10.1007/978-3-030-72869-4_11

Adetuyi, B. O., Adetunji, C. O., Olajide, P. A., Ogunlana, O. O., Mathew, J. T., Inobeme, A., Popoola, O. A., Olaitan, F. Y., Akinbo, O., Oyewole, O. A., Eniola, K. I. T., & Yerima, M. B. (2024). Removal of Microplastic from Wastewater Treatment Plants. In book: Microplastic Pollution. Springer, Singapore, 1, 271–286. DOI: DOI: 10.1007/978-981-99-8357-5_16

Almeida, C. M. R., Merillas, B., & Pontinha, A. D. R. (2024). Trends on Aerogel-Based Biosensors for Medical Applications: An Overview. *International Journal of Molecular Sciences*, *25*(2), 1309. DOI: 10.3390/ijms25021309 PMID: 38279307

Bakhori, N. M., Ismail, Z., Hassan, M. Z., & Dolah, R. (2023). Emerging Trends in Nanotechnology: Aerogel-Based Materials for Biomedical Applications. *Nanomaterials (Basel, Switzerland), 13*(6), 1063. DOI: 10.3390/nano13061063 PMID: 36985957

Bakhori, N. M., Ismail, Z., Hassan, M. Z., & Dolah, R. (2023). Emerging Trends in Nanotechnology: Aerogel-Based Materials for Biomedical Applications. *Nanomaterials (Basel, Switzerland), 13*(6), 1063. DOI: 10.3390/nano13061063 PMID: 36985957

Catauro, M., & Ciprioti, S. V. (2021). Characterization of Hybrid Materials Prepared by Sol-Gel Method for Biomedical Implementations. A Critical Review. *Materials (Basel), 14*(7), 1788. DOI: 10.3390/ma14071788 PMID: 33916333

Chen, Z., Fu, X., Liu, R., Song, Y., & Yin, X. (2023). Fabrication, Performance, and Potential Applications of MXene Composite Aerogels. *Nanomaterials (Basel, Switzerland), 13*(14), 2048. DOI: 10.3390/nano13142048 PMID: 37513059

Dolez, P. I., Marsha, S., & McQueen, R. H. (2022). Fibers and Textiles for Personal Protective Equipment: Review of Recent Progress and Perspectives on Future Developments. *Textiles, 2*(2), 349–381. DOI: 10.3390/textiles2020020

Fan, K., Lin, L., Li, D., Wang, F., & Li, J. (2024). Progress in Research on the Preparation of Graphene-Based Aerogels Using γ-ray Irradiation Technology. *Gels (Basel, Switzerland), 10*(2), 90. DOI: 10.3390/gels10020090 PMID: 38391420

Ferreira-Gonçalves, T., Iglesias-Mejuto, A., Linhares, T., Coelho, J. M. P., Vieira, P., Faísca, P., Catarino, J., Pinto, P., Ferreira, D., Ferreira, H. A., Gaspar, M. M., Durães, L., García-González, C. A., & Reis, C. P. (2022). Biological Thermal Performance of Organic and Inorganic Aerogels as Patches for Photothermal Therapy. *Gels (Basel, Switzerland), 8*(8), 485. DOI: 10.3390/gels8080485 PMID: 36005086

Finny, A. S. (2024). 3D bioprinting in bioremediation: A comprehensive review of principles, applications, and future directions. *PeerJ, 12*, e16897. DOI: 10.7717/peerj.16897 PMID: 38344299

Franco, P., Cardea, S., Tabernero, A., & De Marco, I. (2021). Porous Aerogels and Adsorption of Pollutants from Water and Air: A Review. *Molecules (Basel, Switzerland), 26*(15), 4440. DOI: 10.3390/molecules26154440 PMID: 34361593

García-González, C. A., Budtova, T., Durães, L., Erkey, C., Del Gaudio, P., Gurikov, P., Koebel, M., Liebner, F., Neagu, M., & Smirnova, I. (2019). An Opinion Paper on Aerogels for Biomedical and Environmental Applications. *Molecules (Basel, Switzerland), 24*(9), 1815. DOI: 10.3390/molecules24091815 PMID: 31083427

Greszta, A., Bartkowiak, G., Dąbrowska, A., Gliścińska, E., Machnowski, W., & Kozikowski, P. (2022). Multilayer Nonwoven Inserts with Aerogel/PCMs for the Improvement of Thermophysiological Comfort in Protective Clothing against the Cold. *Materials (Basel), 15*(6), 2307. DOI: 10.3390/ma15062307 PMID: 35329760

Guastaferro, M., Baldino, L., Reverchon, E., & Cardea, S. (2021). Production of Porous Agarose-Based Structures: Freeze-Drying vs. Supercritical CO_2 Drying. *Gels (Basel, Switzerland), 7*(4), 198. DOI: 10.3390/gels7040198 PMID: 34842697

Guastaferro, M., Reverchon, E., & Baldino, L. (2021). Polysaccharide-Based Aerogel Production for Biomedical Applications: A Comparative Review. *Materials (Basel), 14*(7), 1631. DOI: 10.3390/ma14071631 PMID: 33810582

Inobeme, A., Adetunji, C. O., Mathew, J T., Ajai, A. I., Inobeme, J., Bamigboye, M. O., Onyeaku, S., Maliki, M., Eziukwu, C. &Tawa, K. (2023). Nanotechnology for Bioremediation of Heavy Metals. In book: Microbial Technologies in Industrial Wastewater Treatment. Springer Nature Singapore, 19-30. .DOI: 10.1007/978-981-99-2435-6_2

Inobeme, A., Mathew, J.T., Adetunji, C.O., Agbugui, M.O., Osarenren, E.J., Oti, C., Ngonso, B.F., Inobeme, J., Oyedolapo, M.B., Bernard, E. &Jibrin, N.A. (2024). Biocomposites with graphene derivatives. In book: Advances in Biocomposites and their Applications, Elsevier Inc., 149-166. DOI: DOI: 10.1016/B978-0-443-19074-2.00005-8

Jin, R., Zhou, Z., Liu, J., Shi, B., Zhou, N., Wang, X., Jia, X., Guo, D., & Xu, B. (2023). Aerogels for Thermal Protection and Their Application in Aerospace. *Gels (Basel, Switzerland)*, *9*(8), 606. DOI: 10.3390/gels9080606 PMID: 37623061

Khan, N. R., Sharmin, T., & Bin Rashid, A. (2023). Exploring the Versatility of Aerogels: Broad Applications in Biomedical Engineering, Astronautics, Energy Storage, Biosensing, and Current Progress. *Heliyon*, *10*(1), e23102. DOI: 10.1016/j.heliyon.2023.e23102 PMID: 38163169

Khan, N. R., Sharmin, T., & Bin Rashid, A. (2023). Exploring the Versatility of Aerogels: Broad Applications in Biomedical Engineering, Astronautics, Energy Storage, Biosensing, and Current Progress. *Heliyon*, *10*(1), e23102. DOI: 10.1016/j.heliyon.2023.e23102 PMID: 38163169

Lee, K. H., Arshad, Z., Dahshan, A., Alshareef, M., Alsulami, Q. A., Bibi, A., Lee, E.-J., Nawaz, M., Zubair, U., & Javid, A. (2023). Porous Aerogel Structures as Promising Materials for Photocatalysis, Thermal Insulation Textiles, and Technical Applications: A Review. *Catalysts*, *13*(9), 1286. DOI: 10.3390/catal13091286

Lee, K. H., Arshad, Z., Dahshan, A., Alshareef, M., Alsulami, Q. A., Bibi, A., Lee, E.-J., Nawaz, M., Zubair, U., & Javid, A. (2023). Porous Aerogel Structures as Promising Materials for Photocatalysis, Thermal Insulation Textiles, and Technical Applications: A Review. *Catalysts*, *13*(9), 1286. DOI: 10.3390/catal13091286

Li, H., Wang, L., Wei, Y., Yan, W., & Feng, J. (2022). Preparation of Templated Materials and Their Application to Typical Pollutants in Wastewater: A Review. *Frontiers in Chemistry*, *10*, 882876. DOI: 10.3389/fchem.2022.882876 PMID: 35480393

Li, P., Yang, C., Xu, X., Miao, C., He, T., Jiang, B., & Wu, W. (2022). Preparation of Bio-Based Aerogel and Its Adsorption Properties for Organic Dyes. *Gels (Basel, Switzerland)*, *8*(11), 755. DOI: 10.3390/gels8110755 PMID: 36421576

Liu, J., Buahom, P., Lu, C., Yu, H., & Park, C. B. (2022). Microscopic revelation of the solid-gas coupling and Knudsen effect on the thermal conductivity of silica aerogel with inter-connected pores. *Scientific Reports*, *12*(1), 21034. DOI: 10.1038/s41598-022-24133-5 PMID: 36470925

Mathew, J. T., Adetunji, C. O., Inobeme, A., Musah, M., Azeh, Y., Otori, A. A., Shaba, E. Y., Mamman, A., & Tanko, E. (2023). Removal of Heavy Metals Using Bio-remedial Techniques. Springer Nature Switzerland AG 2023 M. P. Shah (ed.), Modern Approaches in Waste Bioremediation, 117-130. DOI: 10.1007/978-3-031-24086-7_6

Mathew, J. T., Inobeme, A., Abdulkadir, A., Etsuyankpa, M. B., Shaba, E. Y., Musah, M., Azeh, Y., Adetunji, C. O., Ibrahim, A. M., Tanko, E., & Mamman, A. (2024). a. Properties and Characteristics of Nanocoatings for Medicinal Applications. In book: Sustainable Approach to Protective Nanocoatings. IGI Global; publishing Tomorrow's. *Advances in Chemical and Materials Engineering*, •••, 203–227. DOI: 10.4018/979-8-3693-3136-1.ch007

Mathew, J. T., Inobeme, A., Musah, M., Azeh, Y., Abdullahi, A., Shaba, E. Y., Salihu, A. M., Muhammad, E. B., Josiah, J. G., Jibrin, N. A., Ismail, H., Muhammad, A. I., Maurice, J., Mamman, A., & Ndamitso, M. M. (2024). b. A Critical Review of Green Approach on Wastewater Treatment Strategies. *Journal of Applied Science & Environmental Management*, *28*(2), 363–391. DOI: 10.4314/jasem.v28i2.9

Mathew, J. T., Musah, M., Azeh, Y., & Muhammed, M. (2023). a. Kinetic Study of Heavy Metals Removal from Pharmaceutical Wastewater Using Geopolymer/Fe_3O_4Nanocomposite. *Bima Journal of Science and Technology*, *7*(4), 152–163. DOI: 10.56892/bima.v7i4.539

Mathew, J. T., Musah, M., Azeh, Y., & Muhammed, M. (2023). b. Adsorptive Removal of Selected Toxic Metals from Pharmaceutical Wastewater using Fe_3O_4/ZnONanocomposite. *Dutse Journal of Pure and Applied Sciences*, *9*(4a), 236–248. DOI: 10.4314/dujopas.v9i4a.22

Nita, L. E., Ghilan, A., Rusu, A. G., Neamtu, I., & Chiriac, A. P. (2020). New Trends in Bio-Based Aerogels. *Pharmaceutics*, *12*(5), 449. DOI: 10.3390/pharmaceutics12050449 PMID: 32414217

Noman, M. T., Amor, N., Ali, A., Petrik, S., Coufal, R., Adach, K., & Fijalkowski, M. (2021). Aerogels for Biomedical, Energy and Sensing Applications. *Gels (Basel, Switzerland)*, *7*(4), 264. DOI: 10.3390/gels7040264 PMID: 34940324

Olorunsola, A., Adetunji, C. O., Olaniyan, O., Adetunji, J. B., Akinbo, K., Adetuyi, B. O., Inobeme, A., Mathew, J. T., & Ajenifujah-Solebo, S. O. (2024). Application of Nanotechnology for the Production of Biopesticides, Bioinsecticides, Bioherbicides, Mosquitoe Repellants and Biofungicides. In book: Handbook of Agricultural Biotechnology, John Wiley & Sons, 293-309. DOI: DOI: 10.1002/9781119836483.ch11

Payanda, K. O., Alsuhile, A. A. A. M., Yousefzadeh, H., Ulker, Z., Bozbag, S. E., García-González, C. A., Smirnova, I., & Erkey, C. (2023). The effect of synthesis conditions and process parameters on aerogel properties. *Frontiers in Chemistry*, *11*, 1294520. DOI: 10.3389/fchem.2023.1294520 PMID: 37937209

Romero-Montero, A., Valencia-Bermúdez, J. L., Rosas-Meléndez, S. A., Núñez-Tapia, I., Piña-Barba, M. C., Leyva-Gómez, G., & Del Prado-Audelo, M. L. (2023). Biopolymeric Fibrous Aerogels: The Sustainable Alternative for Water Remediation. *Polymers*, *15*(2), 262. DOI: 10.3390/polym15020262 PMID: 36679143

Saleh, W. M., Ahmad, M. I., & Yahya, E. B. (2023). Nanostructured Bioaerogels as a Potential Solution for Particulate Matter Pollution. *Gels (Basel, Switzerland)*, *9*, 575. DOI: 10.3390/gels9070575 PMID: 37504454

Sambucci, M., Savoni, F., & Valente, M. (2023). Aerogel Technology for Thermal Insulation of Cryogenic Tanks-Numerical Analysis for Comparison with Traditional Insulating Materials. *Gels (Basel, Switzerland)*, *9*(4), 307. DOI: 10.3390/gels9040307 PMID: 37102919

Seesaard, T., & Wongchoosuk, C. (2023). Flexible and Stretchable Pressure Sensors: From Basic Principles to State-of-the-Art Applications. *Micromachines*, *14*(8), 1638. DOI: 10.3390/mi14081638 PMID: 37630177

Sena-Torralba, A., Álvarez-Diduk, R., Parolo, C., Piper, A., & Merkoçi, A. (2022). Toward Next Generation Lateral Flow Assays: Integration of Nanomaterials. *Chemical Reviews, 122*(18), 14881–14910. DOI: 10.1021/acs.chemrev.1c01012 PMID: 36067039

Shao, Z., Chen, H., Wang, Q., Kang, G., Wang, X., Li, W., Liu, Y., & Zheng, G. (2022). High-performance multifunctional electrospun fibrous air filter for personal protection: A review. *Separation and Purification Technology*, *302*, 122175. DOI: 10.1016/j.seppur.2022.122175 PMID: 36168392

Sheng, Z., Liu, Z., Hou, Y., Jiang, H., Li, Y., Li, G., & Zhang, X. (2023). The Rising Aerogel Fibers: Status, Challenges, and Opportunities. *Advanced Science (Weinheim, Baden-Wurttemberg, Germany)*, *10*(9), e2205762. DOI: 10.1002/advs.202205762 PMID: 36658735

Tawiah, B., Ofori, E. A., & Bin, F. (2023). Scientometric Review of Sustainable Fire-Resistant Polysaccharide-Based Composite Aerogels. *Sustainability (Basel)*, *15*(16), 12185. DOI: 10.3390/su151612185

Thang, N. H., Chien, T. B., & Cuong, D. X. (2023). Polymer-Based Hydrogels Applied in Drug Delivery: An Overview. *Gels (Basel, Switzerland)*, *9*(7), 523. DOI: 10.3390/gels9070523 PMID: 37504402

Yang, D., Xu, P., Tian, C., Li, S., Xing, T., Li, Z., Wang, X., & Dai, P. (2023). Biomass-Derived Flexible Carbon Architectures as Self-Supporting Electrodes for Energy Storage. *Molecules (Basel, Switzerland)*, *28*(17), 6377. DOI: 10.3390/molecules28176377 PMID: 37687208

Zhan, W., Chen, L., Kong, Q., Li, L., Chen, M., Jiang, J., Li, W., Shi, F., & Xu, Z. (2023). The Synthesis and Polymer-Reinforced Mechanical Properties of SiO_2 Aerogels: A Review. *Molecules (Basel, Switzerland)*, *28*(14), 5534. DOI: 10.3390/molecules28145534 PMID: 37513406

Chapter 9
Physicochemical Properties and Types of Aerogels

Abel Inobeme
Edo State University, Uzairue, Nigeria

Chetan Chauhan
https://orcid.org/0000-0002-0515-4733
Sardar Patel University, Mandi, India

Muniratu Maliki
Edo State University, Uzairue, Nigeria

Mathew Adekoya
https://orcid.org/0000-0002-6404-8615
Xi'an Jiaotong University, China

ABSTRACT

Aerogels are a class of materials known for their exceptional physical properties, characterized by their high porosity and low density. They are composed of a network of interconnected nanoparticles or polymers, resulting in a solid with a porous structure filled with gas. This unique structure imparts several advantageous physicochemical properties to

DOI: 10.4018/979-8-3693-5971-6.ch009

aerogels, including high surface area, low thermal conductivity, and high porosity. These properties make aerogels promising candidates for various applications, such as insulation, catalysis, drug delivery, and environmental remediation. This chapter focuses on the physicochemical properties and types of aerogels.

3.1 INTRODUCTION

Aerogels represent a fascinating class of materials known for their remarkable combination of physicochemical properties and unique structural characteristics. These materials have garnered significant interest across various scientific and industrial fields due to their exceptional properties, including high porosity, low density, large surface area, and intriguing optical properties. Understanding the physicochemical properties and types of aerogels is crucial for harnessing their full potential in diverse applications ranging from thermal insulation to drug delivery systems (Bakhori et al. 2023).

One of the defining features of aerogels is their extraordinarily high porosity, typically ranging from 80% to 99.8%. This high porosity is a result of the interconnected network of nanoparticles or polymers that form the aerogel structure. The large void spaces within the aerogel matrix contribute to its low density and exceptional thermal insulation properties. Aerogels are renowned for their extremely low density, often referred to as "frozen smoke" or "solid smoke." This low density is a consequence of their high porosity and minimal solid content, making them some of the lightest solid materials known to humankind. Despite their low density, aerogels can exhibit remarkable mechanical strength and structural integrity (Yang et al. 2022).

The nanoporous structure of aerogels results in an exceptionally high surface area per unit mass. This large surface area facilitates various surface interactions, such as adsorption, catalysis, and sensing applications. The high surface area also enables efficient mass transfer processes, making aerogels suitable for applications requiring rapid diffusion or reaction kinetics. Aerogels possess outstanding thermal insulating properties, primarily due to their low thermal conductivity. The combination of high porosity and low solid content minimizes the material's ability to conduct heat, making aerogels effective thermal insulators. This property has led to their widespread use in applications requiring thermal insulation, such as building materials, protective clothing, and aerospace components (Franco et al. 2021).

Despite their lightweight and porous nature, aerogels can exhibit impressive mechanical strength and rigidity. The interconnected network of nanoparticles or polymers provides structural support, allowing aerogels to withstand mechanical stresses and deformations. This mechanical resilience makes aerogels suitable for use in structural components, reinforcement materials, and impact-resistant coatings. Certain types of aerogels, particularly silica-based aerogels, exhibit exceptional transparency, making them optically intriguing materials. Transparent aerogels have found applications in optical devices, windows, and light-guiding systems, where their combination of low density and optical clarity offers unique advantages (Khan et al. 2023).

In addition, aerogels represent a versatile class of materials with extraordinary physicochemical properties and diverse types tailored for specific applications. Their unique combination of high porosity, low density, and exceptional thermal insulating properties make them promising candidates for a wide range of technological and industrial applications. Understanding the properties and types of aerogels is essential for harnessing their full potential and advancing innovations in various fields (Nita et al. 2020).

3.2 PHYSICOCHEMICAL PROPERTIES OF AEROGELS

Aerogels stand out as an intriguing category of materials, marked by a distinctive blend of physicochemical characteristics that render them immensely appealing across diverse applications. Grasping these properties is paramount to fully unlocking the potential of aerogels across multiple domains, spanning aerospace, energy, environmental remediation, and biomedical fields (Harun-Ur-Rashid et al. 2023).

High Porosity: Aerogels exhibit a defining trait in their remarkably elevated porosity, typically falling within the impressive range of 80% to 99.8%. This exceptional porosity stems from their nanoporous architecture, wherein interconnected networks of nanoparticles or polymers amalgamate to craft a solid matrix adorned with myriad void spaces. The substantial volume of pores serves as the cornerstone for several advantageous properties, chief among them being low density, extensive surface area, and augmented mass transport capabilities. The substantial porosity of aerogels is a consequence of their intricate internal structure, characterized by interconnected voids at nanometer scales. This intricate network facilitates the creation of a solid material with a vast number of empty spaces, resulting in an incredibly high porosity. Such a porous structure is highly advantageous in various applications due to its unique properties. First and foremost, the exceptionally high porosity of aerogels contributes to their low density, making them some of the lightest solid materials known. This property is particularly advantageous in applications where weight reduction is critical, such as aerospace components and lightweight structural materials (Yamasaki et al. 2019; Inobeme et al., 2024).

Furthermore, the extensive surface area provided by the large pore volume of aerogels is beneficial for numerous applications requiring surface interactions, such as adsorption, catalysis, and sensing. The high surface area enhances the material's reactivity and adsorption capacity, making aerogels valuable in environmental remediation, chemical catalysis, and gas storage applications. Additionally, the interconnected pore structure of aerogels facilitates enhanced mass transport capabilities, enabling efficient diffusion of gases and liquids within the material. This property is advantageous in applications such as filtration, drug delivery, and energy storage devices, where rapid mass transport is essential (Noman et al. 2021; Adetunji et al., 2023a).

Low Density: Aerogels boast an extraordinary characteristic in their exceptionally low density, often dubbed as "solid smoke" or "frozen smoke." Thanks to their high porosity and minimal solid content, aerogels rank among the lightest solid materials identified, showcasing densities plummeting to as low as 0.001 g/cm^3. This defining attribute positions aerogels as prime candidates for endeavors where weight minimization holds paramount importance. For instance, in aerospace engineering, where every ounce counts, aerogels find utility as lightweight structural materials for crafting components crucial to flight operations. Their featherweight nature not only aids in reducing the overall weight of aircraft but also contributes to enhancing fuel efficiency and maneuverability (Borella et al. 2023).

Moreover, in the realm of space exploration missions, where payload capacity is severely restricted, aerogels emerge as indispensable assets. Their incredibly low density translates to reduced mass, which in turn facilitates launching payloads into space more economically. Additionally, aerogels excel as insulating materials, shielding spacecraft and equipment from the extreme temperatures prevalent in the vacuum of space. By providing effective thermal insulation while minimizing bulk, aerogels optimize the efficiency of space exploration missions, ensuring the safety and functionality of vital systems and instrumentation. In essence, the unparalleled lightweight nature of aerogels epitomizes their significance in applications necessitating weight reduction, such as aerospace engineering and space exploration. As technological advancements continue to unfold, the versatility and utility of aerogels are expected to further expand, paving the way for innovative solutions in domains where efficiency and performance are paramount concerns (Jin *et al.* 2023).

High Surface Area: Aerogels are remarkable materials renowned for their nanoporous structure, endowing them with an extraordinarily high surface area per unit mass. This unique feature enables aerogels to engage in diverse surface interactions crucial for applications like adsorption, catalysis, and chemical sensing. In contexts such as pollutant capture and removal, aerogels excel due to their ability to adsorb contaminants effectively, facilitated by their extensive surface area. Similarly, in gas separation processes, the large surface area of aerogels enhances their efficiency in selectively capturing specific gases, contributing to the purification of gas streams. Furthermore, in heterogeneous catalysis, where surface interactions are pivotal for reaction kinetics, aerogels serve as excellent catalyst supports, promoting enhanced activity and selectivity. Their large surface area provides ample active sites for catalytic reactions to occur, thus improving overall reaction efficiency. Consequently, aerogels play a crucial role in various industrial and environmental applications, where maximizing surface area is imperative for achieving optimal performance and functionality (Almeida et al. 2024).

Low Thermal Conductivity: Aerogels are renowned for their exceptional thermal insulating properties, primarily attributed to their remarkably low thermal conductivity. This distinctive characteristic arises from the combination of two key factors: high porosity and low solid content. The nanoporous structure of aerogels, with interconnected pores spanning the material, creates a tortuous path for heat transfer. As a result, thermal energy encounters numerous interfaces between the gas-filled pores and the solid matrix, leading to significant resistance to heat flow. The high porosity of aerogels, often exceeding 90%, further enhances their insulating capabilities by maximizing the air-filled voids within the material. Air is a poor conductor of heat compared to solid materials, and the extensive network of pores in aerogels effectively traps air molecules, inhibiting heat conduction through the material. Additionally, the low solid content of aerogels, typically composed of lightweight materials such as silica or carbon, minimizes the solid-to-solid contact points within the structure, thereby reducing thermal conduction pathways (Jin et al. 2023).

These combined factors make aerogels highly effective thermal insulators, finding applications in various industries. In building construction, aerogel-based insulation materials offer superior thermal performance compared to conventional insulating materials, enabling energy-efficient building designs and reducing heating and cooling costs. In aerospace applications, aerogels provide thermal protection for spacecraft, shielding against the extreme temperature variations encountered during space missions. Furthermore, in cryogenic storage systems, where maintaining ultra-low temperatures is critical, aerogels serve as excellent insulation materials to minimize heat transfer and ensure the integrity of stored cryogenic fluids. In addition, the outstanding thermal insulating properties of aerogels, stemming from their low thermal conductivity and high porosity, make them indispensable in applications requiring efficient heat management and thermal insulation (Arshad et al. 2023).

Mechanical Strength: Despite their lightweight and porous composition, aerogels can display impressive mechanical strength and rigidity, attributes that stem from their unique structural characteristics. The interconnected network of nanoparticles or polymers within aerogels forms a resilient framework that provides substantial structural support. This network effectively distributes applied forces throughout the material, helping aerogels withstand mechanical stresses and prevent deformations. The mechanical strength of aerogels can be tailored to meet specific application needs through careful selection of composition and optimization of fabrication methods. For instance, aerogels made from materials such as silica or carbon can exhibit enhanced mechanical properties due to the inherent strength of these substances. Moreover, controlling factors such as pore size, density, and cross-linking density during the fabrication process allows for further customization of mechanical performance (Khan et al. 2023).

These attributes render aerogels versatile materials suitable for a wide range of applications requiring robust mechanical properties. In structural engineering, aerogels can serve as lightweight yet sturdy structural components, offering a balance between strength and weight savings. Additionally, in protective coatings or composites, aerogels can provide reinforcement and impact resistance while maintaining overall material integrity. Generally, despite their lightweight and porous nature, aerogels possess remarkable mechanical strength and rigidity, thanks to their interconnected nanoparticle or polymer networks. With the ability to be engineered to meet specific performance requirements, aerogels hold promise across various industries as structural materials, protective coatings, and beyond (Lázár et al. 2023).

Transparency: Certain types of aerogels, notably silica-based aerogels, showcase remarkable transparency across both the visible and infrared segments of the electromagnetic spectrum. This optical clarity, when coupled with their low density and outstanding thermal insulating capabilities, renders aerogels highly desirable for a myriad of applications in optics, photonics, and solar energy utilization. Transparent aerogels find extensive use in optical devices such as windows, lenses, and light guides, where their ability to transmit light efficiently allows for improved optical performance. Their transparency enables minimal light scattering or absorption, ensuring high-quality optical transmission across a broad spectrum of wavelengths. This makes them valuable in applications requiring precise light manipulation and transmission, such as in optical sensors and imaging systems (Wang et al. 2020; Mathew et al., 2023a).

Moreover, the combination of optical transparency and excellent thermal insulation makes aerogels ideal candidates for solar energy harvesting technologies. Transparent aerogels can be employed as components in solar thermal collectors, allowing sunlight to penetrate while minimizing heat loss through the device. This enhances the efficiency of solar energy conversion processes by maximizing the absorption of solar radiation while reducing thermal losses, thereby improving overall energy conversion efficiency. Overall, the unique combination of optical transparency, low density, and thermal insulation properties positions aerogels as versatile materials in various fields, including optics, photonics, and solar energy utilization. Their utilization in optical components and solar devices underscores their potential to contribute to advancements in renewable energy technologies and optical engineering (An et al. 2022).

Furthermore, aerogels possess a unique set of physicochemical properties that make them highly desirable materials for a wide range of applications. Their high porosity, low density, high surface area, low thermal conductivity, mechanical strength, and transparency offer numerous opportunities for innovation and advancement in various fields. Continued research and development efforts aimed at understanding and optimizing these properties will further expand the potential applications of aerogels in the future (Franco et al. 2021; Mathew et al., 2024a).

Figure 1. Physicochemical properties of aerogel

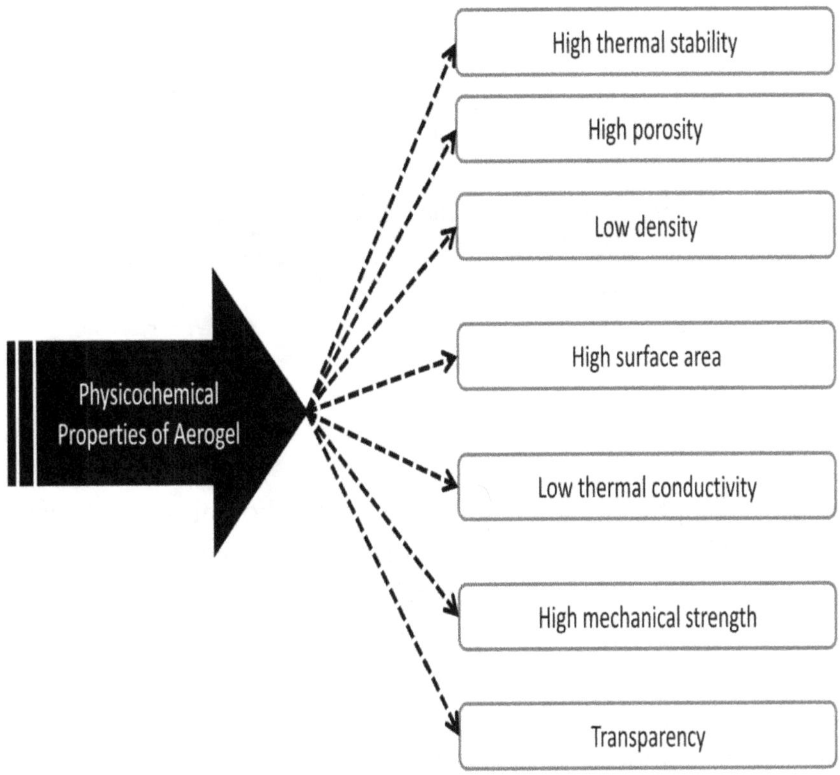

3.3 TYPES OF AEROGELS

There are various types of aerogels (figure 2). The Silica aerogel composed primarily of silicon dioxide (SiO_2), silica aerogels exhibit exceptional thermal insulation, low density, and transparency in both visible and infrared spectra.

Figure 2. Types of aerogel

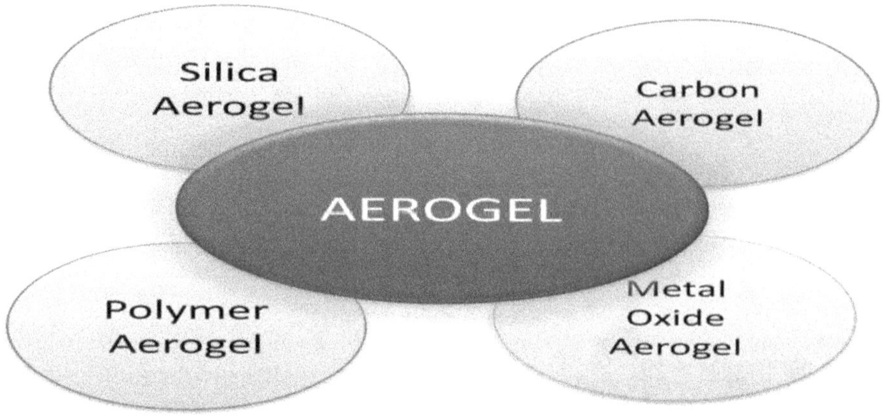

They find applications in insulation, optics, photonics, and solar energy harvesting. Made from carbon-based materials, carbon aerogels are known for their high porosity, lightweight nature, excellent electrical conductivity, and thermal stability. They are utilized in energy storage devices like supercapacitors, as well as in thermal insulation and catalyst supports. These aerogels are constructed from various polymers such as polyurethane and polyimide. They offer flexibility, lightweight characteristics, tunable properties, and good mechanical strength. Polymer aerogels find applications in insulation, packaging materials, and biomedical applications. Comprising metal oxide nanoparticles like titanium dioxide or alumina, metal oxide aerogels boast high surface area, chemical stability, and tunable properties. They are employed in catalysis, gas sensing, and environmental remediation efforts. Each type of aerogel presents distinct advantages and is tailored for specific applications, ranging from thermal insulation to advanced materials in various industries (Ma et al. 2023).

Table 1. Types of aerogels

Type of Aerogel	Composition	Properties	Applications	References
Silica Aerogel	Silicon dioxide (SiO_2)	High surface area, low density, excellent thermal insulation, optical transparency	Insulation, optics, photonics, solar energy harvesting	Wu et al. 2024.
Carbon Aerogel	Carbon-based materials	High porosity, lightweight, excellent electrical conductivity, thermal stability	Energy storage devices (supercapacitors), thermal insulation, catalyst support	Zhang et al. 2022.
Polymer Aerogel	Various polymers (e.g., polyurethane, polyimide)	Flexible, lightweight, tunable properties, good mechanical strength	Insulation, packaging materials, biomedical applications	Bakhoriet al. 2023.
Metal Oxide Aerogel	Metal oxide nanoparticles (e.g., titanium dioxide, alumina)	High surface area, chemical stability, tunable properties	Catalysis, gas sensing, environmental remediation	Rando et al. 2022.

3.4 ENVIRONMENTAL REMEDIATION USING AEROGELS

Environmental remediation involves the removal or mitigation of pollutants and contaminants from soil, water, and air to restore ecological balance and protect human health. Aerogels, with their unique properties, offer innovative solutions for various environmental remediation challenges (Khan et al. 2023).

Oil Spill Cleanup: Aerogels excel in absorbing oil and hydrocarbons from water, thanks to their high surface area and inherent hydrophobicity. When deployed on the water surface, aerogels swiftly adsorb oil while repelling water, facilitating the efficient removal of oil spills. This capability stems from the nanoporous structure of aerogels, which provides ample surface area for oil molecules to adhere to, while their hydrophobic nature prevents water from interfering with the adsorption process (Zamparas et al. 2020; Adetunji et al., 2023b).

Furthermore, aerogels can undergo surface modifications to enhance their affinity for specific types of hydrocarbons. By tailoring the surface chemistry of aerogels, their selectivity for particular hydrocarbons can be optimized, enhancing their effectiveness in oil spill cleanup efforts. This versatility makes aerogels invaluable tools for mitigating the environmental impact of oil spills and facilitating the recovery of contaminated water bodies. In general, aerogels offer a highly efficient and adaptable solution for oil spill cleanup, leveraging their high surface area, hydrophobic nature, and tunable surface chemistry. By rapidly adsorbing oil while repelling water and selectively targeting specific hydrocarbons, aerogels contribute to the restoration of affected ecosystems and the preservation of aquatic environments (Lamy-Mendes et al. 2019; Inobeme et al., 2023).

Heavy Metal Remediation: Metal oxide aerogels, including titanium dioxide or iron oxide nanoparticles, demonstrate remarkable adsorption capabilities for heavy metals. These aerogels have the ability to selectively bind to metal ions present in aqueous solutions, employing chemical interactions or ion exchange processes. This selectivity enables them to efficiently capture heavy metal pollutants, effectively reducing their concentration in water sources. By incorporating metal oxide aerogels into filtration systems or deploying them directly into contaminated water sources, heavy metal pollutants can be efficiently removed. In filtration systems, water passes through the aerogel-based filters, allowing metal ions to be adsorbed onto the surface of the aerogels as the water flows through. Alternatively, metal oxide aerogels can be dispersed directly into contaminated water sources, where they act as sorbents, trapping heavy metal contaminants and preventing their further dispersion. The application of metal oxide aerogels in water treatment processes offers a sustainable approach to reducing environmental toxicity and safeguarding ecosystems and human health. By selectively targeting heavy metal pollutants, these aerogels contribute to the restoration and preservation of water quality, ensuring the availability of clean and safe drinking water for communities worldwide (Damiri et al. 2022).

Water Purification: Silica aerogels, when functionalized with specific chemical groups, offer a versatile solution for targeting a wide range of contaminants in water, including pesticides, pharmaceuticals, and organic pollutants. These functionalized aerogels leverage chemical interactions to selectively adsorb pollutants from water, effectively removing them from the aqueous phase. This process provides an efficient means of water purification, contributing to the removal of harmful substances and the enhancement of water quality (Mathew et al., 2023a; Mathew et al., 2023b).

Moreover, the high porosity of silica aerogels enables rapid filtration of water. The interconnected nanoporous structure of aerogels facilitates the flow of water through the material, allowing for efficient capture and retention of contaminants. This rapid filtration capability enhances the overall efficiency of water treatment processes, enabling faster treatment rates and higher throughput. By integrating aerogel-based adsorbents into water treatment infrastructure, access to clean and safe drinking water can be improved, particularly in areas facing contamination challenges. Whether deployed in centralized treatment plants or decentralized point-of-use systems, aerogel-based adsorbents offer a reliable and cost-effective solution for addressing water quality concerns and ensuring the provision of potable water to communities worldwide (Franco et al. 2021; Mathew et al., 2024b).

Air Quality Improvement: Aerogels serve as valuable components in air filtration systems, targeting a range of airborne pollutants including particulate matter, volatile organic compounds (VOCs), and gases. Carbon aerogels, characterized by their high surface area and porosity, excel in adsorbing VOCs and gases due to their enhanced affinity for organic molecules. This capability makes them effective in capturing harmful pollutants emitted from various sources such as household products, building materials, and industrial processes (Mallakpour et al. 2022).

Furthermore, metal oxide aerogels exhibit catalytic properties that facilitate the decomposition of pollutants present in the air. When incorporated into air filtration systems, these aerogels catalyze chemical reactions that convert harmful substances into less toxic or inert compounds. This process contributes to the purification of indoor air, reducing the concentration of pollutants and mitigating health risks associated with exposure to airborne contaminants. By integrating aerogel-based filters into ventilation systems or portable air purifiers, indoor air quality can be significantly improved. These filters effectively trap and neutralize pollutants, ensuring cleaner and healthier indoor environments for occupants. As a result, the use of aerogel-based filtration technology offers a practical and efficient means of reducing the adverse effects of indoor air pollution on human health and well-being (Wang et al. 2023).

Groundwater Remediation: Polymer aerogels present compelling solutions for groundwater remediation owing to their customizable characteristics and exceptional adsorption capabilities. These aerogels can be tailored to selectively capture various contaminants prevalent in groundwater, including chlorinated solvents, pesticides, and industrial chemicals. Through precise engineering, polymer aerogels can be modified to optimize their affinity for specific pollutants, enhancing remediation efficiency. The lightweight and porous nature of polymer aerogels makes them well-suited for deployment in subsurface environments. Their high porosity facilitates the infiltration of groundwater, allowing them to effectively trap and immobilize contaminants as water passes through. This mechanism prevents further migration of pollutants, thereby halting the spread of contamination and contributing to aquifer restoration efforts (Al-Hashimi et al. 2021).

Furthermore, the tunable properties of polymer aerogels enable versatility in application. By adjusting factors such as pore size, surface chemistry, and polymer composition, aerogels can be customized to address the unique challenges posed by different types of groundwater contamination. This adaptability enhances the effectiveness of remediation strategies, allowing for targeted treatment of specific pollutants in diverse hydrogeological settings. The deployment of polymer aerogels in groundwater remediation offers several advantages. Their high adsorption capacity and selective capture capabilities enable efficient removal of contaminants, reducing the risk of groundwater pollution and protecting drinking water supplies. Additionally, their lightweight and deployable nature facilitate cost-effective and minimally invasive remediation approaches, mitigating environmental risks while promoting sustainable aquifer management. Overall, polymer aerogels represent a promising technology for addressing groundwater contamination and safeguarding water resources for future generations (Romero-Montero et al. 2023).

Nanoparticle Remediation: Aerogels functionalized with nanoparticles, particularly zero-valent iron (ZVI) nanoparticles, offer promising solutions for remediating contaminated soils and groundwater. ZVI-functionalized aerogels leverage the unique properties of both aerogels and ZVI nanoparticles to facilitate the reduction and immobilization of hazardous substances present in the environment. Zero-valent iron nanoparticles possess strong reducing capabilities, allowing them to participate in redox reactions with various contaminants. When incorporated into aerogel matrices, ZVI nanoparticles provide a high surface area and reactive sites for catalyzing chemical transformations. In the presence of ZVI-functionalized aerogels, contaminants such as chlorinated solvents and heavy metals undergo reductive reactions, leading to their transformation into less toxic or immobilized forms. For instance, in the process of reductive dechlorination, ZVI reduces chlorinated solvents by donating electrons, resulting in the breakdown of chlorinated compounds into harmless byproducts (Galdames et al. 2020; Mathew et al., 2024a).

Deploying ZVI-functionalized aerogels in contaminated sites enables targeted treatment of polluted environments. These materials can be injected directly into soil or groundwater matrices, allowing for in situ remediation without the need for extensive excavation or disruption of the site. The high surface area and porosity of aerogels enhance the dispersion and retention of ZVI nanoparticles within the contaminated media, maximizing contact with pollutants and promoting efficient remediation. The use of ZVI-functionalized aerogels offers several advantages for environmental remediation. By facilitating the transformation of toxic contaminants into less harmful forms, these materials contribute to the restoration and rehabilitation of impacted environments. Additionally, the targeted and localized treatment provided by aerogel-based remediation approaches minimizes collateral damage to surrounding ecosystems and reduces the overall cost and duration of cleanup efforts. Overall, ZVI-functionalized aerogels represent a promising technology for addressing soil and groundwater contamination, offering effective and sustainable solutions for environmental restoration (Boccia et al. 2023).

Radiation Cleanup: Aerogels doped with specific elements or compounds, such as boron or lanthanides, offer unique capabilities in addressing radioactive contamination. By incorporating these dopants, aerogels can selectively target and capture radioactive isotopes from soil, water, or air. The doping process modifies the surface chemistry of the aerogels, enhancing their affinity for specific radionuclides. This targeted adsorption enables efficient removal of radioactive contaminants, aiding in the cleanup of sites affected by nuclear accidents, uranium mining, or other industrial activities involving radioactive materials (Singh et al. 2023).

Furthermore, aerogels play a crucial role in radiation shielding applications. By integrating aerogels into radiation shielding materials, such as concrete or composite panels, exposure risks to personnel involved in remediation operations can be minimized. Aerogels effectively attenuate ionizing radiation by absorbing and scattering radiation particles, thereby reducing the dose received by individuals working in contaminated environments. Additionally, aerogel-based shielding materials offer advantages such as low density, flexibility, and ease of handling compared to traditional lead-based shielding alternatives. The versatility of aerogels in both capturing radioactive contaminants and providing effective radiation shielding makes them valuable tools in nuclear waste management and remediation efforts. Whether deployed for in situ remediation of contaminated sites or incorporated into protective barriers and equipment, aerogels contribute to mitigating the environmental and health hazards associated with radioactive materials. Their unique properties and tailored functionalities enable targeted and efficient solutions for managing radioactive contamination and ensuring the safety of workers and communities involved in cleanup operations (Xu et al. 2022).

In addition, aerogels offer versatile and effective solutions for environmental remediation across various media, including water, soil, and air. Their unique properties, such as high surface area, porosity, and tunable chemistry, enable targeted removal and mitigation of a wide range of pollutants and contaminants. By harnessing the capabilities of aerogels, sustainable approaches to environmental cleanup can be realized, fostering ecosystem restoration and safeguarding public health (Khan et al. 2023).

3.5 AEROSPACE

The aerospace industry relies on cutting-edge technologies and materials to meet the demanding requirements of space exploration, aviation, and defense. Aerogels, with their unique combination of properties, have emerged as valuable materials in aerospace applications. Here's an overview of how aerogels are utilized in the aerospace sector (Norizan et al. 2022).

Thermal Insulation: Aerogels are renowned for their exceptional thermal insulation properties. In the aerospace industry, where extreme temperatures are encountered, aerogels play a vital role in providing thermal protection to spacecraft, satellites, and aircraft. They are used in insulation blankets, tiles, and coatings to prevent heat transfer and maintain stable internal temperatures. Aerogel-based thermal insulation systems help protect sensitive components from temperature fluctuations during space missions and atmospheric re-entry (Jin et al. 2023).

Lightweight Structural Components: Despite their low density, aerogels can exhibit remarkable mechanical strength and stiffness. This makes them attractive materials for lightweight structural components in aerospace vehicles. Aerogel-based composites are used to construct panels, fairings, and other structural elements, reducing overall weight while maintaining structural integrity. In spacecraft design, where every gram counts, aerogel-based materials offer significant weight savings without compromising performance (Blanco et al. 2021).

Impact Absorption: Aerogels possess excellent shock-absorbing properties, making them ideal materials for impact protection in aerospace applications. In spacecraft and aircraft, aerogel-based cushions or coatings can mitigate the effects of sudden deceleration during landings or collisions. By absorbing and dissipating impact energy, aerogels help reduce structural damage and enhance the safety of astronauts, passengers, and valuable equipment (Wang et al. 2023).

Acoustic Dampening: In aircraft interiors, noise reduction is essential for passenger comfort. Aerogels can serve as effective acoustic dampening materials, reducing noise levels and vibration. By incorporating aerogels into cabin insulation, wall panels, and flooring, aerospace manufacturers can attenuate engine noise and create a quieter and more comfortable cabin environment for passengers and crew.

Space Exploration Equipment: Aerogels play a crucial role in space exploration missions by enabling the development of innovative equipment and technologies. For example, aerogel-based materials are used in the construction of lightweight and durable components for rovers, landers, and scientific instruments deployed on planetary surfaces. Aerogels can also be utilized as dust mitigation coatings, protecting sensitive equipment from abrasive particles encountered during exploration missions (Bakhori et al. 2023).

Thermal Protection Systems (TPS): During atmospheric re-entry, spacecraft experience extreme heat due to air friction. Aerogels are incorporated into thermal protection systems (TPS) to withstand these high temperatures. Aerogel-based TPS materials act as ablative shields, dissipating heat and preventing the spacecraft's structure from overheating. This ensures the safe re-entry and recovery of spacecraft returning from orbit (Jin et al. 2023).

Energy Storage: Aerogels hold promise for energy storage applications in aerospace systems. Carbon aerogels, in particular, are being explored for use in lightweight and high-performance batteries and capacitors. These aerogel-based energy storage devices offer significant advantages in terms of energy density, power output, and durability, making them ideal for powering spacecraft, satellites, and other aerospace vehicles (Goyal et al. 2022).

Aerodynamic Enhancements: Aerogels can be used to improve the aerodynamic performance of aircraft and spacecraft. Their lightweight nature and customizable properties make them ideal materials for shaping airfoils, reducing drag, and improving fuel efficiency. By incorporating aerogel-based components into aerodynamic designs, aerospace engineers can optimize vehicle performance and achieve better maneuverability and stability (Bakhori et al. 2023).

Furthermore, aerogels have emerged as versatile and valuable materials in the aerospace industry, offering a wide range of applications from thermal insulation and structural components to impact absorption and energy storage. Their unique combination of properties, including lightweight, thermal insulation, mechanical strength, and versatility, make them indispensable for space exploration, aviation, and defense applications. As aerospace technology continues to evolve, aerogels are likely to play an increasingly significant role in advancing the capabilities and safety of aerospace vehicles and systems (Jin et al., 2023).

3.6 SYNTHESIS AND FABRICATION METHODS OF AEROGELS

Aerogels, remarkable materials known for their ultra-lightweight and nanoporous structure, can be synthesized through various methods, each offering distinct advantages and applications. Here's an overview of some common synthesis and fabrication methods of aerogels (Khan et al.2023).

Supercritical Drying: In this method, a wet gel undergoes solvent exchange followed by drying under supercritical conditions. The supercritical drying process involves exposing the gel to a fluid above its critical temperature and pressure, allowing the solvent to be removed while maintaining the gel's structure. This results in the formation of an aerogel with a highly porous and uniform structure. Supercritical drying is favored for producing aerogels with low density, high porosity, and minimal shrinkage. However, it requires specialized equipment and is relatively expensive (Şahin et al. 2017).

Sol-Gel Process: The sol-gel process involves the chemical synthesis of aerogels through the formation of a sol followed by gelation. In this method, metal alkoxides or other precursors are hydrolyzed and polymerized to form a colloidal suspension (sol), which undergoes gelation to form a three-dimensional network. The gel is then dried to remove the solvent, resulting in the formation of an aerogel. The sol-gel process offers versatility, allowing for the incorporation of dopants and additives to tailor the aerogel's properties. However, it can be time-consuming, and shrinkage during drying is a common issue (Esposito, 2019).

Freeze Drying: Freeze drying, also known as lyophilization, involves freezing a wet gel followed by sublimation of the solvent under vacuum conditions. This method preserves the gel's porous structure by removing the solvent directly from the solid state, resulting in minimal shrinkage. Freeze drying is particularly suited for producing aerogels with high porosity and low density. However, it is limited to certain solvents and can be a slow process (Nowak and Jakubczyk, 2020).

Ambient Pressure Drying: Ambient pressure drying, also referred to as conventional drying, involves drying the wet gel under atmospheric pressure with solvent exchange. This method typically involves replacing the solvent with a less volatile one before drying. Ambient pressure drying is relatively simple and cost-effective compared to supercritical drying. However, it may result in lower pore volume and reduced porosity compared to other methods (Abebe et al. 2022).

Rapid Supercritical Extraction: Rapid supercritical extraction is a variation of supercritical drying that involves rapid extraction of the solvent using supercritical conditions. In this method, the wet gel is exposed to supercritical fluid, which rapidly removes the solvent while maintaining the gel's structure. Rapid supercritical extraction offers the advantage of shorter processing times and minimal shrinkage compared to conventional supercritical drying. However, it requires specialized equipment and may be more expensive. These synthesis and fabrication methods offer a range of advantages and are selected based on the desired properties and applications of the aerogel. Aerogels produced through these methods find applications across various industries, including aerospace, biomedical, environmental, and energy sectors (Kravanja et al. 2022).

In aerospace, aerogels are utilized for thermal insulation, structural components, impact absorption, acoustic dampening, and space exploration equipment. Their lightweight nature combined with excellent thermal insulation properties makes them ideal materials for spacecraft components, such as walls, windows, and thermal protection systems. Additionally, aerogels can be incorporated into aircraft interiors for acoustic dampening to reduce noise levels and enhance passenger comfort. In space exploration missions, aerogels are used in the construction of lightweight and durable components for rovers, landers, and scientific instruments deployed on planetary surfaces (Khan et al. 2023).

3.7 CHEMICAL VAPOR DEPOSITION (CVD) OF AEROGELS

Chemical Vapor Deposition (CVD) of aerogels involves several steps (Table 2). Firstly, gaseous precursors containing the desired constituents are introduced into a reaction chamber. These precursors undergo decomposition or reaction upon contact with a substrate or seed material. Nucleation occurs as the decomposed precursors form solid particles or clusters on the substrate's surface. Through continued precursor decomposition, these nucleated particles grow, forming a porous aerogel structure. Finally, the synthesized aerogel may undergo post-processing treatments such as drying, annealing, or surface modification to optimize its properties. CVD offers advantages like precise control over precursor composition and tailorable porosity. Challenges include ensuring uniform nucleation and growth across the substrate, as well as compatibility with diverse precursor chemistries. Overall, CVD is a versatile method for synthesizing aerogels with tailored properties for various applications (Bhowmik & GovindRajan, 2022).

Table 2. Chemical vapor deposition (CVD) of aerogels

Step	Description	Advantages	Challenges	References
Precursor Delivery	Gaseous precursors are introduced into a reaction chamber where they undergo decomposition or reaction.	Precise control over precursor composition and flow	Compatibility with diverse precursor chemistries	Hamachi et al. 2019; Adetunji et al., 2021
Nucleation	Decomposed precursors nucleate to form solid particles or clusters on a substrate or seed material.	Allows for controlled growth and morphology	Achieving uniform nucleation across the substrate	Garcia-Gil et al. 2021.
Growth	Nucleated particles grow through continued precursor decomposition, forming a porous aerogel structure.	Tailorable porosity and structure	Maintaining uniform growth and avoiding agglomeration	Sheng et al. 2033.
Post-Processing	The synthesized aerogel may undergo additional treatments such as drying, annealing, or surface modification.	Enables optimization of aerogel properties	Ensuring structural integrity and stability	Bakhori et al. 2023.

3.8 FUTURE PERSPECTIVES AND CHALLENGES OF AEROGELS

Future perspectives for aerogels include exploring new materials and combinations to expand their range of applications. Integration of advanced functionalities such as self-healing, shape memory, and stimuli-responsive behavior offers exciting possibilities. Scaling up production processes and commercializing aerogel-based products are crucial for meeting industrial demand. Additionally, adopting sustainable practices in aerogel synthesis and utilizing eco-friendly precursor materials are essential for environmental sustainability.

Challenges lie in optimizing manufacturing methods for large-scale production while ensuring cost-effectiveness and market acceptance of aerogel products. Overcoming complexities in synthesis and processing, as well as addressing concerns related to energy consumption and waste management, are critical for the widespread adoption of aerogels. Despite these challenges, advancements in material science and nanotechnology offer opportunities to overcome limitations and unlock the full potential of aerogels in various industries, ranging from aerospace and construction to biomedical and environmental applications. As research and development efforts continue, aerogels are poised to play an increasingly significant role in addressing global challenges and driving innovation in materials science (**Table 3**).

Table 3. Future perspectives and challenges of aerogels

Aspect	Description	Future Perspectives	Challenges	References
Material Diversity	Exploration of new materials and combinations to expand the range of aerogel applications.	Development of novel aerogel compositions for specific needs.	Ensuring compatibility and scalability of production methods.	Chai et al. 2022.
Enhanced Functionality	Integration of advanced functionalities such as self-healing, shape memory, and stimuli-responsive behavior.	Advancements in material science and nanotechnology.	Overcoming complexities in synthesis and processing.	Kantaros et al. 2024.
Scale-Up and Commercialization	Scaling up production processes to meet industrial demand and commercializing aerogel-based products.	Optimization of manufacturing methods for large-scale production.	Cost-effectiveness and market acceptance of aerogel products.	Linhares et al. 2023.
Environmental Sustainability	Utilization of sustainable and eco-friendly precursor materials and green synthesis methods.	Adoption of green chemistry principles and sustainable practices.	Addressing concerns related to energy consumption and waste management.	Larrañaga-Tapia et al.2023)

3.9 CONCLUSION

In general, aerogels represent a remarkable class of materials characterized by their unique physicochemical properties and diverse types. Their exceptionally high surface area, low density, and nanoporous structure make them valuable in various applications across industries. Silica aerogels, renowned for their thermal insulation and optical transparency, find use in aerospace, construction, and optics. Carbon aerogels, with their high porosity and electrical conductivity, are ideal for energy storage devices and catalyst supports. Polymer aerogels offer flexibility and tunable properties, making them suitable for biomedical scaffolds and insulation materials.

Metal oxide aerogels exhibit excellent adsorption properties for pollutants and heavy metals, contributing to environmental remediation efforts.

Despite their incredible potential, aerogels face challenges in large-scale production, cost-effectiveness, and environmental sustainability. However, ongoing research and technological advancements continue to address these challenges, paving the way for broader applications and commercialization. As interdisciplinary collaboration and innovation drive the development of novel aerogel compositions and fabrication techniques, the future holds promise for further unlocking the capabilities of aerogels and harnessing their full potential to address global challenges and contribute to advancements in science, technology, and sustainability.

REFERENCES

Abebe, A. M., Soraru, G. D., Thothadri, G., Andoshe, D. M., Zambotti, A., Ahmed, G. M. S., Tirth, V., & Algahtani, A. (2022). Synthesis and Characterization of High Surface Area Transparent SiOC Aerogels from Hybrid Silicon Alkoxide: A Comparison between Ambient Pressure and Supercritical Drying. *Materials (Basel), 15*(4), 1277. DOI: 10.3390/ma15041277 PMID: 35207820

Adetunji, C. O., Egbuna, C., Ajiboye, A. A. A., Ajayi, O. O., Dauda, W. P., Ghazanfar, S., Ogundolie, F. A., Mathew, J. T., Adetunji, J. B., & Inobeme, A., OlotuTitilayo, Olaniyan, O. T., Ijabadeniyi, O. A., Aborode, A. T., Ficai, A., Akram, M., Bodunrinde, R. E., Laila, U., Olaniyan, O. P., Micheal, O. S., Aliyu, A., Akinbo, O. &Adetuyi, B. O. (2023)a. Biogenic nanomaterials with diverse biological activities in the food and biomedical industries, Evaporative Coolers for the Postharvest Management of Fruits and Vegetables, Academic Press, 395-420, . DOI: 10.3390/molecules26195913 PMID: 34641456

Adetunji, C. O., Mathew, J. T., Inobeme, A., Taiwo, O., Olaniyan, O., Shakira, G., & Wadazani, D.(2023)b.Application of Biosensors for Detection and Monitoring of Water Quality.In book: Organic and Inorganic Materials Based Sensors, John Wiley & Sons, Inc., 1065-1077, DOI: 10.3390/molecules26195913 PMID: 34641456

Adetunji, C. O., Olaniyan, O. T., Inobeme, A., Mathew, J. T., & Ahamed, M. I. (2021). Waterborne Polyurethanes for Biomedical Applications. In: Inamuddin, Boddula R., Khan A. (eds) Sustainable Production and Applications of Waterborne Polyurethanes. Advances in Science, Technology & Innovation (IEREK Interdisciplinary Series for Sustainable Development). Springer, Cham. DOI: 10.1007/978-3-030-72869-4_11

Al-Hashimi, O., Hashim, K., Loffill, E., Marolt Čebašek, T., Nakouti, I., Faisal, A. A. H., & Al-Ansari, N. (2021). A Comprehensive Review for Groundwater Contamination and Remediation: Occurrence, Migration and Adsorption Modelling. *Molecules (Basel, Switzerland), 26*(19), 5913. DOI: 10.3390/molecules26195913 PMID: 34641456

Almeida, C. M. R., Merillas, B., & Pontinha, A. D. R. (2024). Trends on Aerogel-Based Biosensors for Medical Applications: An Overview. *International Journal of Molecular Sciences*, *25*(2), 1309. DOI: 10.3390/ijms25021309 PMID: 38279307

An, L., Li, Z., Guo, Z., Hu, Y., Huang, Y., Armstrong, J. N., Zhou, C., & Ren, S. (2022). Transparent thermal insulation ceramic aerogel materials for solar thermal conversion. *Nanoscale Advances*, *4*(20), 4291–4295. DOI: 10.1039/D2NA00412G PMID: 36321153

Arshad, A., Roy, A., Mallick, T. K., & Tahir, A. A. (2023). Shape-Stabilized PEGylated Silica Aerogel-Composite as an Energy Saving Building Material. *Industrial & Engineering Chemistry Research*, *62*(47), 20236–20250. DOI: 10.1021/acs.iecr.3c02373 PMID: 38045733

Bakhori, N. M., Ismail, Z., Hassan, M. Z., & Dolah, R. (2023). Emerging Trends in Nanotechnology: Aerogel-Based Materials for Biomedical Applications. *Nanomaterials (Basel, Switzerland)*, *13*(6), 1063. DOI: 10.3390/nano13061063 PMID: 36985957

Bhowmik, S., & Govind Rajan, A. (2022). Chemical vapor deposition of 2D materials: A review of modeling, simulation, and machine learning studies. *iScience*, *25*(3), 103832. DOI: 10.1016/j.isci.2022.103832 PMID: 35243221

Blanco, D., Rubio, E. M., Lorente-Pedreille, R. M., & Sáenz-Nuño, M. A. (2021). Lightweight Structural Materials in Open Access: Latest Trends. *Materials (Basel)*, *14*(21), 6577. DOI: 10.3390/ma14216577 PMID: 34772103

Boccia, A. C., Neagu, M., & Pulvirenti, A. (2023). Bio-Based Aerogels for the Removal of Heavy Metal Ions and Oils from Water: Novel Solutions for Environmental Remediation. *Gels (Basel, Switzerland)*, *10*(1), 32. DOI: 10.3390/gels10010032 PMID: 38247754

Borella, L., Rozo, A., Perfetti, C., & Iorio, C. S. (2023). Characterization of Composite Freeze-Dried Aerogels with Simulant Lunar Regolith for Space Applications. *Materials (Basel)*, *16*(17), 5797. DOI: 10.3390/ma16175797 PMID: 37687489

Chai, B., Zhang, W., Liu, Y., Zhu, S., Gu, Z., & Zhang, H. (2022). Progress in Research and Application of Graphene Aerogel-A Bibliometric Analysis. *Materials (Basel), 16*(1), 272. DOI: 10.3390/ma16010272 PMID: 36614611

Damiri, F., Andra, S., Kommineni, N., Balu, S. K., Bulusu, R., Boseila, A. A., Akamo, D. O., Ahmad, Z., Khan, F. S., Rahman, M. H., Berrada, M., & Cavalu, S. (2022). Recent Advances in Adsorptive Nanocomposite Membranes for Heavy Metals Ion Removal from Contaminated Water: A Comprehensive Review. *Materials (Basel), 15*(15), 5392. DOI: 10.3390/ma15155392 PMID: 35955327

Esposito, S. (2019). "Traditional" Sol-Gel Chemistry as a Powerful Tool for the Preparation of Supported Metal and Metal Oxide Catalysts. *Materials (Basel), 12*(4), 668. DOI: 10.3390/ma12040668 PMID: 30813441

Franco, P., Cardea, S., Tabernero, A., & De Marco, I. (2021). Porous Aerogels and Adsorption of Pollutants from Water and Air: A Review. *Molecules (Basel, Switzerland), 26*(15), 4440. DOI: 10.3390/molecules26154440 PMID: 34361593

Galdames, A., Ruiz-Rubio, L., Orueta, M., Sánchez-Arzalluz, M., & Vilas-Vilela, J. L. (2020). Zero-Valent Iron Nanoparticles for Soil and Groundwater Remediation. *International Journal of Environmental Research and Public Health, 17*(16), 5817. DOI: 10.3390/ijerph17165817 PMID: 32796749

Garcia-Gil, A., Biswas, S., & Holmes, J. D. (2021). A Review of Self-Seeded Germanium Nanowires: Synthesis, Growth Mechanisms and Potential Applications. *Nanomaterials (Basel, Switzerland), 11*(8), 2002. DOI: 10.3390/nano11082002 PMID: 34443831

Goyal, D., Dang, R. K., Goyal, T., Saxena, K. K., Mohammed, K. A., & Dixit, S. (2022). Graphene: A Path-Breaking Discovery for Energy Storage and Sustainability. *Materials (Basel), 15*(18), 6241. DOI: 10.3390/ma15186241 PMID: 36143552

Hamachi, L. S., Yang, H., Jen-La Plante, I., Saenz, N., Qian, K., Campos, M. P., Cleveland, G. T., Rreza, I., Oza, A., Walravens, W., Chan, E. M., Hens, Z., Crowther, A. C., & Owen, J. S. (2019). Precursor reaction kinetics control compositional grading and size of CdSe1-x S x nanocrystalheterostructures. *Chemical Science (Cambridge)*, *10*(26), 6539–6552. DOI: 10.1039/C9SC00989B PMID: 31367306

Harun-Ur-Rashid, M., Jahan, I., Foyez, T., & Imran, A. B. (2023). Bio-Inspired Nanomaterials for Micro/Nanodevices: A New Era in Biomedical Applications. *Micromachines*, *14*(9), 1786. DOI: 10.3390/mi14091786 PMID: 37763949

Inobeme, A., Adetunji, C. O., Mathew, J T., Ajai, A. I., Inobeme, J., Bamigboye, M. O., Onyeaku, S., Maliki, M., Eziukwu, C. &Tawa, K. (2023). Nanotechnology for Bioremediation of Heavy Metals. In book: Microbial Technologies in Industrial Wastewater Treatment. Springer Nature Singapore, 19-30. .DOI: 10.1007/978-981-99-2435-6_2

Inobeme, A., Mathew, J.T., Adetunji, C.O., Agbugui, M.O., Osarenren, E.J., Oti, C., Ngonso, B.F., Inobeme, J., Oyedolapo, M.B., Bernard, E. &Jibrin, N.A. (2024). Biocomposites with graphene derivatives. In book: Advances in Biocomposites and their Applications, Elsevier Inc., 149-166. DOI: DOI: 10.1016/B978-0-443-19074-2.00005-8

Jin, R., Zhou, Z., Liu, J., Shi, B., Zhou, N., Wang, X., Jia, X., Guo, D., & Xu, B. (2023). Aerogels for Thermal Protection and Their Application in Aerospace. *Gels (Basel, Switzerland)*, *9*(8), 606. DOI: 10.3390/gels9080606 PMID: 37623061

Kantaros, A., Ganetsos, T., & Petrescu, F. I. T. (2024). Transforming Object Design and Creation: Biomaterials and Contemporary Manufacturing Leading the Way. *Biomimetics (Basel, Switzerland)*, *9*(1), 48. DOI: 10.3390/biomimetics9010048 PMID: 38248622

Khan, I. U., Qi, S. S., Gul, F., Manan, S., Rono, J. K., Naz, M., Shi, X. N., Zhang, H., Dai, Z. C., & Du, D. L. (2023). A Green Approach Used for Heavy Metals 'Phytoremediation' Via Invasive Plant Species to Mitigate Environmental Pollution: A Review. *Plants*, *12*(4), 725. DOI: 10.3390/plants12040725 PMID: 36840073

Khan, N. R., Sharmin, T., & Bin Rashid, A. (2023). Exploring the Versatility of Aerogels: Broad Applications in Biomedical Engineering, Astronautics, Energy Storage, Biosensing, and Current Progress. *Heliyon, 10*(1), e23102. DOI: 10.1016/j.heliyon.2023.e23102 PMID: 38163169

Kravanja, K. A., Finšgar, M., Knez, Ž., & Knez Marevci, M. (2022). Supercritical Fluid Technologies for the Incorporation of Synthetic and Natural Active Compounds into Materials for Drug Formulation and Delivery. *Pharmaceutics, 14*(8), 1670. DOI: 10.3390/pharmaceutics14081670 PMID: 36015296

Lamy-Mendes, A., Torres, R. B., Vareda, J. P., Lopes, D., Ferreira, M., Valente, V., Girão, A. V., Valente, A. J. M., & Durães, L. (2019). Amine Modification of Silica Aerogels/Xerogels for Removal of Relevant Environmental Pollutants. *Molecules (Basel, Switzerland), 24*(20), 3701. DOI: 10.3390/molecules24203701 PMID: 31618901

Larrañaga-Tapia, M., Betancourt-Tovar, B., Videa, M., Antunes-Ricardo, M., & Cholula-Díaz, J. L. (2023). Green synthesis trends and potential applications of bimetallic nanoparticles towards the sustainable development goals 2030. *Nanoscale Advances, 6*(1), 51–71. DOI: 10.1039/D3NA00761H PMID: 38125589

Lázár, I., Čelko, L., & Menelaou, M. (2023). Aerogel-Based Materials in Bone and Cartilage Tissue Engineering-A Review with Future Implications. *Gels (Basel, Switzerland), 9*(9), 746. DOI: 10.3390/gels9090746 PMID: 37754427

Linhares, T., Carneiro, V. H., Pessoa de Amorim, M. T., & Durães, L. (2023). A Comparative Thermoacoustic Insulation Study of Silica Aerogels Reinforced with Reclaimed Textile Fibres: Cotton, Polyester and Wool. *Gels (Basel, Switzerland), 9*(7), 548. DOI: 10.3390/gels9070548 PMID: 37504426

Ma, B., Cheng, Y., Hu, P., Fang, D., & Wang, J. (2023). Passive Daytime Radiative Cooling of Silica Aerogels. *Nanomaterials (Basel, Switzerland), 13*(3), 467. DOI: 10.3390/nano13030467 PMID: 36770428

Mallakpour, S., Azadi, E., & Hussain, C. M. (2022). Fabrication of air filters with advanced filtration performance for removal of viral aerosols and control the spread of COVID-19. *Advances in Colloid and Interface Science, 303*, 102653. DOI: 10.1016/j.cis.2022.102653 PMID: 35349924

Mathew, J. T., Inobeme, A., Abdulkadir, A., Etsuyankpa, M. B., Shaba, E. Y., Musah, M., Azeh, Y., Adetunji, C. O., Ibrahim, A. M., Tanko, E., & Mamman, A. (2024). a. Properties and Characteristics of Nanocoatings for Medicinal Applications. In book: Sustainable Approach to Protective Nanocoatings. IGI Global; publishing Tomorrow's. *Advances in Chemical and Materials Engineering*, •••, 203–227. DOI: 10.4018/979-8-3693-3136-1.ch007

Mathew, J. T., Inobeme, A., Musah, M., Azeh, Y., Abdullahi, A., Shaba, E. Y., Salihu, A. M., Muhammad, E. B., Josiah, J. G., Jibrin, N. A., Ismail, H., Muhammad, A. I., Maurice, J., Mamman, A., & Ndamitso, M. M. (2024). b. A Critical Review of Green Approach on Wastewater Treatment Strategies. *Journal of Applied Science & Environmental Management, 28*(2), 363–391. DOI: 10.4314/jasem.v28i2.9

Mathew, J. T., Inobeme, A., Musah, M., Azeh, Y., Abdullahi, A., Shaba, E. Y., Salihu, A. M., Muhammad, E. B., Josiah, J. G., Jibrin, N. A., Ismail, H., Muhammad, A. I., Maurice, J., Mamman, A., & Ndamitso, M. M. (2024). A Critical Review of Green Approach on Wastewater Treatment Strategies. *Journal of Applied Science & Environmental Management, 28*(2), 363–391. DOI: 10.4314/jasem.v28i2.9

Mathew, J. T., Musah, M., Azeh, Y., & Muhammed, M. (2023). a. Adsorptive Removal of Selected Toxic Metals from Pharmaceutical Wastewater using Fe_3O_4/ZnONanocomposite. *Dutse Journal of Pure and Applied Sciences, 9*(4a), 236–248. DOI: 10.4314/dujopas.v9i4a.22

Mathew, J. T., Musah, M., Azeh, Y., & Muhammed, M. (2023). b. Kinetic Study of Heavy Metals Removal from Pharmaceutical Wastewater Using Geopolymer/Fe_3O_4Nanocomposite. *Bima Journal of Science and Technology, 7*(4), 152–163. DOI: 10.56892/bima.v7i4.539

Nita, L. E., Ghilan, A., Rusu, A. G., Neamtu, I., & Chiriac, A. P. (2020). New Trends in Bio-Based Aerogels. *Pharmaceutics*, *12*(5), 449. DOI: 10.3390/pharmaceutics12050449 PMID: 32414217

Noman, M. T., Amor, N., Ali, A., Petrik, S., Coufal, R., Adach, K., & Fijalkowski, M. (2021). Aerogels for Biomedical, Energy and Sensing Applications. *Gels (Basel, Switzerland)*, *7*(4), 264. DOI: 10.3390/gels7040264 PMID: 34940324

Norizan, M. N., Shazleen, S. S., Alias, A. H., Sabaruddin, F. A., Asyraf, M. R. M., Zainudin, E. S., Abdullah, N., Samsudin, M. S., Kamarudin, S. H., & Norrrahim, M. N. F. (2022). Nanocellulose-Based Nanocomposites for Sustainable Applications: A Review. *Nanomaterials (Basel, Switzerland)*, *12*(19), 3483. DOI: 10.3390/nano12193483 PMID: 36234612

Nowak, D., & Jakubczyk, E. (2020). The Freeze-Drying of Foods-The Characteristic of the Process Course and the Effect of Its Parameters on the Physical Properties of Food Materials. *Foods*, *9*(10), 1488. DOI: 10.3390/foods9101488 PMID: 33080983

Rando, G., Sfameni, S., Galletta, M., Drommi, D., Cappello, S., & Plutino, M. R. (2022). Functional Nanohybrids and Nanocomposites Development for the Removal of Environmental Pollutants and Bioremediation. *Molecules (Basel, Switzerland)*, *27*(15), 4856. DOI: 10.3390/molecules27154856 PMID: 35956804

Romero-Montero, A., Valencia-Bermúdez, J. L., Rosas-Meléndez, S. A., Núñez-Tapia, I., Piña-Barba, M. C., Leyva-Gómez, G., & Del Prado-Audelo, M. L. (2023). Biopolymeric Fibrous Aerogels: The Sustainable Alternative for Water Remediation. *Polymers*, *15*(2), 262. DOI: 10.3390/polym15020262 PMID: 36679143

Şahin, İ., Özbakır, Y., İnönü, Z., Ulker, Z., & Erkey, C. (2017). Kinetics of Supercritical Drying of Gels. *Gels (Basel, Switzerland)*, *4*(1), 3. DOI: 10.3390/gels4010003 PMID: 30674780

Sheng, Z., Liu, Z., Hou, Y., Jiang, H., Li, Y., Li, G., & Zhang, X. (2023). The Rising Aerogel Fibers: Status, Challenges, and Opportunities. Advanced science (Weinheim, Baden-Wurttemberg, Germany), 10(9), e2205762. DOI: 10.1002/advs.202205762

Singh, G., Bhadange, S., Bhawna, F., Shewale, P., Dahiya, R., Aggarwal, A., Manju, F., & Arya, S. K. (2023). Phytoremediation of radioactive elements, possibilities and challenges: Special focus on agricultural aspects. *International Journal of Phytoremediation, 25*(1), 1–8. DOI: 10.1080/15226514.2022.2043239 PMID: 35244498

Wang, J., Petit, D., & Ren, S. (2020). Transparent thermal insulation silica aerogels. *Nanoscale Advances*, *2*(12), 5504–5515. DOI: 10.1039/D0NA00655F PMID: 36133881

Wang, M., Zhu, P., Liu, S., Chen, Y., Liang, D., Liu, Y., Chen, W., Du, L., & Wu, C. (2023). Application of Nanozymes in Environmental Monitoring, Management, and Protection. *Biosensors (Basel), 13*(3), 314. DOI: 10.3390/bios13030314 PMID: 36979526

Wu, H., Shang, D., Zhang, H., Zhi, L., Sun, S., Cui, S., & Yan, C. (2024). Phase-transition materials derived photonic metamaterials for passively dynamic solar thermal and coldness harvesting. *Heliyon, 10*(2), e23986. DOI: 10.1016/j.heliyon.2024.e23986 PMID: 38293359

Xu, H., Li, X., Tong, Z., Zhang, B., & Ji, H. (2022). Thermal Radiation Shielding and Mechanical Strengthening of Mullite Fiber/SiC Nanowire Aerogels Using In Situ Synthesized SiC Nanowires. *Materials (Basel), 15*(10), 3522. DOI: 10.3390/ma15103522 PMID: 35629551

Yamasaki, S., Sakuma, W., Yasui, H., Daicho, K., Saito, T., Fujisawa, S., Isogai, A., & Kanamori, K. (2019). NanocelluloseXerogelsWith High Porosities and Large Specific Surface Areas. *Frontiers in Chemistry, 7*, 316. DOI: 10.3389/fchem.2019.00316 PMID: 31134187

Yang, X., Jiang, P., Xiao, R., Fu, R., Liu, Y., Ji, C., Song, Q., Miao, C., Yu, H., Gu, J., Wang, Y., & Sai, H. (2022). Robust Silica-Agarose Composite Aerogels with Interpenetrating Network Structure by In Situ Sol-Gel Process. *Gels (Basel, Switzerland), 8*(5), 303. DOI: 10.3390/gels8050303 PMID: 35621601

Zamparas, M., Tzivras, D., Dracopoulos, V., & Ioannides, T. (2020). Application of Sorbents for Oil Spill Cleanup Focusing on Natural-Based Modified Materials: A Review. *Molecules (Basel, Switzerland), 25*(19), 4522. DOI: 10.3390/molecules25194522 PMID: 33023166

Zhang, L., Lei, Y., He, P., Wu, H., Guo, L., & Wei, G. (2022). Carbon Material-Based Aerogels for Gas Adsorption: Fabrication, Structure Design, Functional Tailoring, and Applications. *Nanomaterials (Basel, Switzerland)*, *12*(18), 3172. DOI: 10.3390/nano12183172 PMID: 36144967

Compilation of References

Abbas, Q., Shinde, P. A., Abdelkareem, M. A., Alami, A. H., Mirzaeian, M., Yadav, A., & Olabi, A. G. (2022). Graphene Synthesis Techniques and Environmental Applications. *Materials (Basel), 15*(21), 7804. DOI: 10.3390/ma15217804 PMID: 36363396

Abdul Khalil, H. P. S., Adnan, A. S., Yahya, E. B., Olaiya, N. G., Safrida, S., Hossain, M. S., Balakrishnan, V., Gopakumar, D. A., Abdullah, C. K., Oyekanmi, A. A., & Pasquini, D. (2020). A Review on Plant Cellulose Nanofibre-Based Aerogels for Biomedical Applications. *Polymers, 12*(8), 1759. DOI: 10.3390/polym12081759 PMID: 32781602

Abdul-Khalil, H. P. S., Jha, K., Yahya, E. B., Panchal, S., Patel, N., Garai, A., Kumari, S., & Jameel, M. (2023). Insights into the Potential of Biopolymeric Aerogels as an Advanced Soil-Fertilizer Delivery Systems. *Gels (Basel, Switzerland), 9*(8), 666. DOI: 10.3390/gels9080666 PMID: 37623121

Abdulmalek, S., Nasir, A., Jabbar, W. A., Almuhaya, M. A. M., Bairagi, A. K., Khan, M. A., & Kee, S. H. (2022). IoT-Based Healthcare-Monitoring System towards Improving Quality of Life: A Review. *Healthcare (Basel), 10*(10), 1993. DOI: 10.3390/healthcare10101993 PMID: 36292441

Abebe, A. M., Soraru, G. D., Thothadri, G., Andoshe, D. M., Zambotti, A., Ahmed, G. M. S., Tirth, V., & Algahtani, A. (2022). Synthesis and Characterization of High Surface Area Transparent SiOC Aerogels from Hybrid Silicon Alkoxide: A Comparison between Ambient Pressure and Supercritical Drying. *Materials (Basel), 15*(4), 1277. DOI: 10.3390/ma15041277 PMID: 35207820

Adetunji, C. O., Egbuna, C., Ajiboye, A. A. A., Ajayi, O. O., Dauda, W. P., Ghazanfar, S., Ogundolie, F. A., Mathew, J. T., Adetunji, J. B., Inobeme, A., OlotuTitilayo, Olaniyan, O. T., Ijabadeniyi, O. A., Aborode, A. T., Ficai, A., Akram, M., Bodunrinde, R. E., Laila, U., Olaniyan, O. P., Micheal, O. S., Aliyu, A., Akinbo, O. &Adetuyi, B. O. (2023)a. Biogenic nanomaterials with diverse biological activities in the food and biomedical industries, Evaporative Coolers for the Postharvest Management of Fruits and Vegetables, Academic Press, 395-420, . DOI: 10.1007/978-981-99-8357-5_16

Adetunji, C. O., Olaniyan, O. T., Inobeme, A., Mathew, J. T., & Ahamed, M. I. (2021). Waterborne Polyurethanes for Biomedical Applications. In: Inamuddin, Boddula R., Khan A. (eds) Sustainable Production and Applications of Waterborne Polyurethanes. Advances in Science, Technology & Innovation (IEREK Interdisciplinary Series for Sustainable Development). Springer, Cham. DOI: 10.1007/978-3-030-72869-4_11

Adetunji, C. O., Egbuna, C., Ajiboye, A. A. A., Ajayi, O. O., Dauda, W. P., Ghazanfar, S., Ogundolie, F. A., Mathew, J. T., Adetunji, J. B., & Inobeme, A., OlotuTitilayo, Olaniyan, O. T., Ijabadeniyi, O. A., Aborode, A. T., Ficai, A., Akram, M., Bodunrinde, R. E., Laila, U., Olaniyan, O. P., Micheal, O. S., Aliyu, A., Akinbo, O. &Adetuyi, B. O. (2023)a. Biogenic nanomaterials with diverse biological activities in the food and biomedical industries, Evaporative Coolers for the Postharvest Management of Fruits and Vegetables, Academic Press, 395-420, . DOI: 10.1002/btm2.10124 PMID: 30680322

Adetunji, C. O., Egbuna, C., Ajiboye, A. A. A., Ajayi, O. O., Dauda, W. P., Ghazanfar, S., Ogundolie, F. A., Mathew, J. T., Adetunji, J. B., & Inobeme, A., OlotuTitilayo, Olaniyan, O. T., Ijabadeniyi, O. A., Aborode, A. T., Ficai, A., Akram, M., Bodunrinde, R. E., Laila, U., Olaniyan, O. P., Micheal, O. S., Aliyu, A., Akinbo, O. &Adetuyi, B. O. (2023)a. Biogenic nanomaterials with diverse biological activities in the food and biomedical industries, Evaporative Coolers for the Postharvest Management of Fruits and Vegetables, Academic Press, 395-420, . DOI: 10.3390/molecules26195913 PMID: 34641456

Adetunji, C. O., Mathew, J. T., Inobeme, A., Taiwo, O., Olaniyan, O., Shakira, G., & Wadazani, D. (2024). Application of Biosensors for Detection and Monitoring of Water Quality. In book: Organic and Inorganic Materials Based Sensors, John Wiley & Sons, Inc., 1065-1077, DOI: 10.1371/journal.pone.0247608 PMID: 33770082

Adetuyi, B. O., Mathew, J. T., Inobeme, A., Falana, Y. O., Adetunji, C. O., Shahnawaz, M., Oyewole, O. A., Eniola, K. I. T., Yerima, M. B., & Popoola, O. A. (2024). Tyres, Bitumen Wear and Plastic Bottles, Other Single Used Plastic as Major Sources of Microplastic. In book: Microplastic Pollution. Springer, Singapore, 1, 141–160. DOI: DOI: 10.1007/978-981-99-8357-5_9

Ahmad, A., & Kamaruddin, M. A., H P S, A. K., Yahya, E. B., Muhammad, S., Rizal, S., Ahmad, M. I., Surya, I., & Abdullah, C. K. (. (2023). Recent Advances in Nanocellulose Aerogels for Efficient Heavy Metal and Dye Removal. *Gels (Basel, Switzerland)*, *9*(5), 416. DOI: 10.3390/gels9050416 PMID: 37233007

Ahmad, F., Salem-Bekhit, M. M., Khan, F., Alshehri, S., Khan, A., Ghoneim, M. M., Wu, H. F., Taha, E. I., & Elbagory, I. (2022). Unique Properties of Surface-Functionalized Nanoparticles for Bio-Application: Functionalization Mechanisms and Importance in Application. *Nanomaterials (Basel, Switzerland)*, *12*(8), 1333. DOI: 10.3390/nano12081333 PMID: 35458041

Ahmed, A. T. M. F., Islam, M. Z., Mahmud, M. S., Sarker, M. E., & Islam, M. R. (2022). Hemp as a potential raw material toward a sustainable world: A review. *Heliyon, 8*(1), e08753. DOI: 10.1016/j.heliyon.2022.e08753 PMID: 35146149

Al Soubaihi, R. M., Saoud, K. M., Awadallah-F, A., Elkhatat, A. M., Al-Muhtaseb, S. A., & Dutta, J. (2023). Investigation of palladium catalysts in mesoporous silica support for CO oxidation and CO2 adsorption. *Heliyon, 9*(7), e18354. DOI: 10.1016/j.heliyon.2023.e18354 PMID: 37539214

Alahi, M. E. E., Sukkuea, A., Tina, F. W., Nag, A., Kurdthongmee, W., Suwannarat, K., & Mukhopadhyay, S. C. (2023). Integration of IoT-Enabled Technologies and Artificial Intelligence (AI) for Smart City Scenario: Recent Advancements and Future Trends. *Sensors (Basel), 23*(11), 5206. DOI: 10.3390/s23115206 PMID: 37299934

Almeida, C. M. R., Merillas, B., & Pontinha, A. D. R. (2024). Trends on Aerogel-Based Biosensors for Medical Applications: An Overview. *International Journal of Molecular Sciences, 25*(2), 1309. DOI: 10.3390/ijms25021309 PMID: 38279307

Alshangiti, D. M., El-Damhougy, T. K., Zaher, A., Madani, M., & Mohamady ghobashy, M. (2023). Revolutionizing biomedicine: advancements, applications, and prospects of nanocomposite macromolecular carbohydrate-based hydrogel biomaterials: a review. *RSC Advances, 13*(50), 35251–35291. DOI: 10.1039/D3RA07391B PMID: 38053691

Amaraweera, S. M., Gunathilake, C. A., Gunawardene, O. H. P., Dassanayake, R. S., Cho, E. B., & Du, Y. (2023). Carbon Capture Using Porous Silica Materials. *Nanomaterials (Basel, Switzerland), 13*(14), 2050. DOI: 10.3390/nano13142050 PMID: 37513061

Ang, L. Y. L., & Cui, F. (2022). Remote work: Aircraft noise implications, prediction, and management in the built environment. Applied acoustics. Acoustique applique. *Applied Acoustics, 198*, 108978. DOI: 10.1016/j.apacoust.2022.108978 PMID: 36034578

Anil Kumar, Y., Koyyada, G., Ramachandran, T., Kim, J. H., Sajid, S., Moniruzzaman, M., Alzahmi, S., & Obaidat, I. M. (2023). Carbon Materials as a Conductive Skeleton for Supercapacitor Electrode Applications: A Review. *Nanomaterials (Basel, Switzerland)*, *13*(6), 1049. DOI: 10.3390/nano13061049 PMID: 36985942

Ankley, G. T., Cureton, P., Hoke, R. A., Houde, M., Kumar, A., Kurias, J., Lanno, R., McCarthy, C., Newsted, J., Salice, C. J., Sample, B. E., Sepúlveda, M. S., Steevens, J., & Valsecchi, S. (2021). Assessing the Ecological Risks of Per- and Polyfluoroalkyl Substances: Current State-of-the Science and a Proposed Path Forward. *Environmental Toxicology and Chemistry*, *40*(3), 564–605. DOI: 10.1002/etc.4869 PMID: 32897586

An, L., Guo, Z., Li, Z., Fu, Y., Hu, Y., Huang, Y., Yao, F., Zhou, C., & Ren, S. (2022). Tailoring thermal insulation architectures from additive manufacturing. *Nature Communications*, *13*(1), 4309. DOI: 10.1038/s41467-022-32027-3 PMID: 35879371

An, L., Li, Z., Guo, Z., Hu, Y., Huang, Y., Armstrong, J. N., Zhou, C., & Ren, S. (2022). Transparent thermal insulation ceramic aerogel materials for solar thermal conversion. *Nanoscale Advances*, *4*(20), 4291–4295. DOI: 10.1039/D2NA00412G PMID: 36321153

Annu, M., Mittal, M., Tripathi, S., & Shin, D. K. (2024). Biopoly-mericNanocomposites for Wastewater Remediation: An Overview on Recent Progress and Challenges. *Polymers*, *16*(2), 294. DOI: 10.3390/polym16020294 PMID: 38276702

Anžlovar, A., & Žagar, E. (2022). Cellulose Structures as a Support or Template for Inorganic Nanostructures and Their Assemblies. *Nanomaterials (Basel, Switzerland)*, *12*(11), 1837. DOI: 10.3390/nano12111837 PMID: 35683693

Arshad, A., Roy, A., Mallick, T. K., & Tahir, A. A. (2023). Shape-Stabilized PEGylated Silica Aerogel-Composite as an Energy Saving Building Material. *Industrial & Engineering Chemistry Research*, *62*(47), 20236–20250. DOI: 10.1021/acs.iecr.3c02373 PMID: 38045733

Asghar, N., Hussain, A., Nguyen, D. A., Ali, S., Hussain, I., Junejo, A., & Ali, A. (2024). Advancement in nanomaterials for environmental pollutants remediation: A systematic review on bibliometrics analysis, material types, synthesis pathways, and related mechanisms. *Journal of Nanobiotechnology*, *22*(1), 26. DOI: 10.1186/s12951-023-02151-3 PMID: 38200605

Ashammakhi, N., GhavamiNejad, A., Tutar, R., Fricker, A., Roy, I., Chatzistavrou, X., Hoque Apu, E., Nguyen, K.-L., Ahsan, T., Pountos, I., & Caterson, E. J. (2022). Highlights on Advancing Frontiers in Tissue Engineering. *Tissue Engineering. Part B, Reviews*, *28*(3), 633–664. DOI: 10.1089/ten.teb.2021.0012 PMID: 34210148

Asmussen, R. M., Turner, J., Chong, S., & Riley, B. J. (2022). Review of recent developments in iodine wasteform production. *Frontiers in Chemistry*, *10*, 1043653. DOI: 10.3389/fchem.2022.1043653 PMID: 36618856

Azeh, Y., Musah, M., Mathew, J. T., Badeggi, U. M., Estuyankpa, M. B., Muhammad, Z. D., Oluwaseyi, D. S., Umar, M. T., Abdulkarim, M. A., Muhammad, A. I., Ibrahim, A. A., Ismail, H., Muhammad, K. T., Umar, G., & Ahmed, T. (2023). In Situ Synthesis of Polyaniline Nanohybrid and Formulation of Polyaniline/Carboxymethyl cellulose/Ethylene Glycol Nanocomposite: Study of its Conducting and Antibacterial Properties. *Journal of Materials and Polymer Science*, *3*(4), 1–9. DOI: 10.47485/2832-9384.1041

Aziz, T., Farid, A., Haq, F., Kiran, M., Ullah, A., Zhang, K., Li, C., Ghazanfar, S., Sun, H., Ullah, R., Ali, A., Muzammal, M., Shah, M., Akhtar, N., Selim, S., Hagagy, N., Samy, M., & Al Jaouni, S. K. (2022). A Review on the Modification of Cellulose and Its Applications. *Polymers*, *14*(15), 3206. DOI: 10.3390/polym14153206 PMID: 35956720

Babcock-Jackson, L., Konovalova, T., Krogman, J. P., Bird, R., & Díaz, L. L. (2023). Sustainable Fertilizers: Publication Landscape on Wastes as Nutrient Sources, Wastewater Treatment Processes for Nutrient Recovery, Biorefineries, and Green Ammonia Synthesis. *Journal of Agricultural and Food Chemistry*, *71*(22), 8265–8296. DOI: 10.1021/acs.jafc.3c00454 PMID: 37219570

Bagshaw, M., &Illig, P. (2019). The Aircraft Cabin Environment. Travel Medicine, 429–436. DOI: 10.1016/B978-0-323-54696-6.00047-1

Bakhori, N. M., Ismail, Z., Hassan, M. Z., & Dolah, R. (2023). Emerging Trends in Nanotechnology: Aerogel-Based Materials for Biomedical Applications. *Nanomaterials (Basel, Switzerland), 13*(6), 1063. DOI: 10.3390/nano13061063 PMID: 36985957

Baranwal, J., Barse, B., Gatto, G., Broncova, G., & Kumar, A. (2022). Electrochemical Sensors and Their Applications: A Review. *Chemosensors (Basel, Switzerland), 10*(9), 363. DOI: 10.3390/chemosensors10090363

Bari, G. A. K. M. R., & Jeong, J. H. (2024). Comprehensive Insights and Advancements in Gel Catalysts for Electrochemical Energy Conversion. *Gels (Basel, Switzerland), 10*(1), 63. DOI: 10.3390/gels10010063 PMID: 38247786

Barnes, M., Sulé-Suso, J., Millett, J., & Roach, P. (2023). Fourier transform infrared spectroscopy as a non-destructive method for analysing herbarium specimens. *Biology Letters, 19*(3), 20220546. DOI: 10.1098/rsbl.2022.0546 PMID: 36946131

Barrios, E., Fox, D., Li Sip, Y. Y., Catarata, R., Calderon, J. E., Azim, N., Afrin, S., Zhang, Z., & Zhai, L. (2019). Nanomaterials in Advanced, High-Performance Aerogel Composites: A Review. *Polymers, 11*(4), 726. DOI: 10.3390/polym11040726 PMID: 31010008

Bashir, I., Lone, F. A., Bhat, R. A., Mir, S. A., Dar, Z. A., & Dar, S. A. (2020). Concerns and Threats of Contamination on Aquatic Ecosystems. Bioremediation and Biotechnology: Sustainable Approaches to Pollution Degradation, 1–26. DOI: 10.1007/978-3-030-35691-0_1

Basso, C. R., Crulhas, B. P., Castro, G. R., & Pedrosa, V. A. (2023). Recent Advances in Functional Nanomaterials for Diagnostic and Sensing Using Self-Assembled Monolayers. *International Journal of Molecular Sciences, 24*(13), 10819. DOI: 10.3390/ijms241310819 PMID: 37445998

Batisha, A. (2022). Horizon scanning process to foresight emerging issues in Arabsphere's water vision. *Scientific Reports, 12*(1), 12709. DOI: 10.1038/s41598-022-16803-1 PMID: 35882917

Batko, K., & Ślęzak, A. (2022). The use of Big Data Analytics in healthcare. *Journal of Big Data, 9*(1), 3. DOI: 10.1186/s40537-021-00553-4 PMID: 35013701

Batool, F., Özçelik, H., Stutz, C., Gegout, P. Y., Benkirane-Jessel, N., Petit, C., & Huck, O. (2021). Modulation of immune-inflammatory responses through surface modifications of biomaterials to promote bone healing and regeneration. *Journal of Tissue Engineering, 12*, 20417314211041428. DOI: 10.1177/20417314211041428 PMID: 34721831

Bhattacharjee, S., Joshi, R., Chughtai, A. A., & Macintyre, C. R. (2019). Graphene Modified Multifunctional Personal Protective Clothing. *Advanced Materials Interfaces, 6*(21), 1900622. DOI: 10.1002/admi.201900622 PMID: 32313805

Bhowmik, S., & Govind Rajan, A. (2022). Chemical vapor deposition of 2D materials: A review of modeling, simulation, and machine learning studies. *iScience, 25*(3), 103832. DOI: 10.1016/j.isci.2022.103832 PMID: 35243221

Blanco, D., Rubio, E. M., Lorente-Pedreille, R. M., & Sáenz-Nuño, M. A. (2021). Lightweight Structural Materials in Open Access: Latest Trends. *Materials (Basel), 14*(21), 6577. DOI: 10.3390/ma14216577 PMID: 34772103

Boccia, A. C., Neagu, M., & Pulvirenti, A. (2023). Bio-Based Aerogels for the Removal of Heavy Metal Ions and Oils from Water: Novel Solutions for Environmental Remediation. *Gels (Basel, Switzerland), 10*(1), 32. DOI: 10.3390/gels10010032 PMID: 38247754

Boccia, A. C., Pulvirenti, A., García-González, C. A., Grisi, F., & Neagu, M. (2023). Compendium of Safety Regulatory for Safe Applications of Aerogels. *Gels (Basel, Switzerland), 9*(11), 842. DOI: 10.3390/gels9110842 PMID: 37998932

Bontempi, E., Sorrentino, G. P., Zanoletti, A., Alessandri, I., Depero, L. E., & Caneschi, A. (2021). Sustainable Materials and their Contribution to the Sustainable Development Goals (SDGs): A Critical Review Based on an Italian Example. *Molecules (Basel, Switzerland), 26*(5), 1407. DOI: 10.3390/molecules26051407 PMID: 33807763

Borella, L., Rozo, A., Perfetti, C., & Iorio, C. S. (2023). Characterization of Composite Freeze-Dried Aerogels with Simulant Lunar Regolith for Space Applications. *Materials (Basel), 16*(17), 5797. DOI: 10.3390/ma16175797 PMID: 37687489

Bouchaut, B., Asveld, L., Hanefeld, U., & Vlierboom, A. (2021). Value Conflicts in Designing for Safety: Distinguishing Applications of Safe-by-Design and the Inherent Safety Principles. *International Journal of Environmental Research and Public Health, 18*(4), 1963. DOI: 10.3390/ijerph18041963 PMID: 33670483

Budtova, T., Aguilera, D. A., Beluns, S., Berglund, L., Chartier, C., Espinosa, E., Gaidukovs, S., Klimek-Kopyra, A., Kmita, A., Lachowicz, D., Liebner, F., Platnieks, O., Rodríguez, A., Navarro, L. K. T., Zou, F., & Buwalda, S. J. (2020). Biorefinery Approach for Aerogels. *Polymers, 12*(12), 2779. DOI: 10.3390/polym12122779 PMID: 33255498

Buratti, C., Moretti, E., Belloni, E., & Agosti, F. (2014). Development of Innovative Aerogel Based Plasters: Preliminary Thermal and Acoustic Performance Evaluation. *Sustainability (Basel), 6*(9), 5839–5852. DOI: 10.3390/su6095839

Burdon, J., Budnik, L. T., Baur, X., Hageman, G., Howard, C. V., Roig, J., Coxon, L., Furlong, C. E., Gee, D., Loraine, T., Terry, A. V.Jr, Midavaine, J., Petersen, H., Bron, D., Soskolne, C. L., & Michaelis, S. (2023). Health consequences of exposure to aircraft contaminated air and fume events: A narrative review and medical protocol for the investigation of exposed aircrew and passengers. *Environmental Health : A Global Access Science Source, 22*(1), 43. DOI: 10.1186/s12940-023-00987-8 PMID: 37194087

Butt, M. A., Voronkov, G. S., Grakhova, E. P., Kutluyarov, R. V., Kazanskiy, N. L., & Khonina, S. N. (2022). Environmental Monitoring: A Comprehensive Review on Optical Waveguide and Fiber-Based Sensors. *Biosensors (Basel), 12*(11), 1038. DOI: 10.3390/bios12111038 PMID: 36421155

Caddeo, F., Loche, D., Casula, M. F., & Corrias, A. (2020). Growing CeO2 Nanoparticles Within the Nano-Porous Architecture of the SiO2 Aerogel. *Frontiers in Chemistry, 8*, 57. DOI: 10.3389/fchem.2020.00057 PMID: 32117882

Cao, B., Xu, J., Wang, F., Zhang, Y., & O'Connor, D. (2021). Vertical Barriers for Land Contamination Containment: A Review. *International Journal of Environmental Research and Public Health, 18*(23), 12643. DOI: 10.3390/ijerph182312643 PMID: 34886373

Carriero, V. C., Di Muzio, L., Petralito, S., Casadei, M. A., & Paolicelli, P. (2023). Cryogel Scaffolds for Tissue-Engineering: Advances and Challenges for Effective Bone and Cartilage Regeneration. *Gels (Basel, Switzerland), 9*(12), 979. DOI: 10.3390/gels9120979 PMID: 38131965

Catauro, M., & Ciprioti, S. V. (2021). Characterization of Hybrid Materials Prepared by Sol-Gel Method for Biomedical Implementations. A Critical Review. *Materials (Basel), 14*(7), 1788. DOI: 10.3390/ma14071788 PMID: 33916333

Chai, B., Zhang, W., Liu, Y., Zhu, S., Gu, Z., & Zhang, H. (2022). Progress in Research and Application of Graphene Aerogel-A Bibliometric Analysis. *Materials (Basel), 16*(1), 272. DOI: 10.3390/ma16010272 PMID: 36614611

Chen, Z., Fu, X., Liu, R., Song, Y., & Yin, X. (2023). Fabrication, Performance, and Potential Applications of MXene Composite Aerogels. *Nanomaterials (Basel, Switzerland), 13*(14), 2048. DOI: 10.3390/nano13142048 PMID: 37513059

Chughtai, A. A., & Khan, W. (2020). Use of personal protective equipment to protect against respiratory infections in Pakistan: A systematic review. *Journal of Infection and Public Health, 13*(3), 385–390. DOI: 10.1016/j.jiph.2020.02.032 PMID: 32146139

Cristea, M., Ionita, D., & Iftime, M. M. (2020). Dynamic Mechanical Analysis Investigations of PLA-Based Renewable Materials: How Are They Useful? *Materials (Basel), 13*(22), 5302. DOI: 10.3390/ma13225302 PMID: 33238537

Czerwinski, F. (2021). Current Trends in Automotive Lightweighting Strategies and Materials. *Materials (Basel), 14*(21), 6631. DOI: 10.3390/ma14216631 PMID: 34772154

Damiri, F., Andra, S., Kommineni, N., Balu, S. K., Bulusu, R., Boseila, A. A., Akamo, D. O., Ahmad, Z., Khan, F. S., Rahman, M. H., Berrada, M., & Cavalu, S. (2022). Recent Advances in Adsorptive Nanocomposite Membranes for Heavy Metals Ion Removal from Contaminated Water: A Comprehensive Review. *Materials (Basel), 15*(15), 5392. DOI: 10.3390/ma15155392 PMID: 35955327

Dolez, P. I., Marsha, S., & McQueen, R. H. (2022). Fibers and Textiles for Personal Protective Equipment: Review of Recent Progress and Perspectives on Future Developments. *Textiles, 2*(2), 349–381. DOI: 10.3390/textiles2020020

Dong, Y., Kong, J., Mousavi, S., Rismanchi, B., & Yap, P.-S. (2023). Wall Insulation Materials in Different Climate Zones: A Review on Challenges and Opportunities of Available Alternatives. *Thermo, 3*(1), 38–65. DOI: 10.3390/thermo3010003

Drdova, S., Zhao, S., Giannakou, M., Sivaraman, D., Guerrero-Alburquerque, N., Bonnin, A., Pauer, R., Pan, Z., Billeter, E., Siqueira, G., Zeng, Z., Koebel, M. M., Malfait, W. J., & Wang, J. (2022). Biomimetic Light-Driven Aerogel Passive Pump for Volatile Organic Pollutant Removal. Advanced science (Weinheim, Baden-Wurttemberg, Germany), 9(11), e2105819. DOI: 10.1002/advs.202105819

Efremenko, E., Stepanov, N., Senko, O., Maslova, O., Lyagin, I., & Aslanli, A. (2023). Progressive Biocatalysts for the Treatment of Aqueous Systems Containing Pharmaceutical Pollutants. *Life (Basel, Switzerland), 13*(3), 841. DOI: 10.3390/life13030841 PMID: 36983996

Eifert, T., Eisen, K., Maiwald, M., & Herwig, C. (2020). Current and future requirements to industrial analytical infrastructure-part 2: Smart sensors. *Analytical and Bioanalytical Chemistry, 412*(9), 2037–2045. DOI: 10.1007/s00216-020-02421-1 PMID: 32055909

Elhenawy, S., Khraisheh, M., AlMomani, F., Hassan, M. K., Al-Ghouti, M. A., & Selvaraj, R. (2021). Recent Developments and Advancements in Graphene-Based Technologies for Oil Spill Cleanup and Oil-Water Separation Processes. *Nanomaterials (Basel, Switzerland), 12*(1), 87. DOI: 10.3390/nano12010087 PMID: 35010035

Esposito, S. (2019). "Traditional" Sol-Gel Chemistry as a Powerful Tool for the Preparation of Supported Metal and Metal Oxide Catalysts. *Materials (Basel), 12*(4), 668. DOI: 10.3390/ma12040668 PMID: 30813441

Fan, K., Lin, L., Li, D., Wang, F., & Li, J. (2024). Progress in Research on the Preparation of Graphene-Based Aerogels Using γ-ray Irradiation Technology. *Gels (Basel, Switzerland), 10*(2), 90. DOI: 10.3390/gels10020090 PMID: 38391420

Farpón, M. G., Peláez, R., Recio, V., Atakan, B., Zaldo, C., & Prieto, G. (2023). Multifunctional materials for catalyst-specific heating and thermometry in tandem catalysis. *Journal of Materials Chemistry. A, Materials for Energy and Sustainability, 11*(37), 19854–19859. DOI: 10.1039/D3TA03654E PMID: 38013847

Fascista, A. (2022). Toward Integrated Large-Scale Environmental Monitoring Using WSN/UAV/Crowdsensing: A Review of Applications, Signal Processing, and Future Perspectives. *Sensors (Basel), 22*(5), 1824. DOI: 10.3390/s22051824 PMID: 35270970

Fediuk, R., Amran, M., Vatin, N., Vasilev, Y., Lesovik, V., & Ozbakkaloglu, T. (2021). Acoustic Properties of Innovative Concretes: A Review. *Materials (Basel), 14*(2), 398. DOI: 10.3390/ma14020398 PMID: 33466943

Ferreira-Gonçalves, T., Constantin, C., Neagu, M., Reis, C. P., Sabri, F., &Simón-Vázquez, R. (2021). Safety and efficacy assessment of aerogels for biomedical applications. *Biomedicine & pharmacotherapy = Biomedecine&pharmacotherapie, 144*, 112356. DOI: 10.1016/j.biopha.2021.112356

Ferreira-Gonçalves, T., Iglesias-Mejuto, A., Linhares, T., Coelho, J. M. P., Vieira, P., Faísca, P., Catarino, J., Pinto, P., Ferreira, D., Ferreira, H. A., Gaspar, M. M., Durães, L., García-González, C. A., & Reis, C. P. (2022). Biological Thermal Performance of Organic and Inorganic Aerogels as Patches for Photothermal Therapy. *Gels (Basel, Switzerland)*, *8*(8), 485. DOI: 10.3390/gels8080485 PMID: 36005086

Fijalkowski, M., Ali, A., Qamer, S., Coufal, R., Adach, K., & Petrik, S. (2023). Hybrid and Single-Component Flexible Aerogels for Biomedical Applications: A Review. *Gels (Basel, Switzerland)*, *10*(1), 4. DOI: 10.3390/gels10010004 PMID: 38275842

Fijalkowski, M., Coufal, R., Ali, A., Adach, K., Petrik, S., Bu, H., & Karl, C. W. (2023). Flexible Hybrid and Single-Component Aerogels: Synthesis, Characterization, and Applications. *Langmuir*, *39*(47), 16760–16775. DOI: 10.1021/acs.langmuir.3c01811 PMID: 37955990

Filipovic, L., & Grasser, T. (2022). Special Issue on Miniaturized Transistors, Volume II. *Micromachines*, *13*(4), 603. DOI: 10.3390/mi13040603 PMID: 35457908

Finny, A. S. (2024). 3D bioprinting in bioremediation: A comprehensive review of principles, applications, and future directions. *PeerJ*, *12*, e16897. DOI: 10.7717/peerj.16897 PMID: 38344299

Flynn, M. A., Check, P., Steege, A. L., Sivén, J. M., & Syron, L. N. (2021). Health Equity and a Paradigm Shift in Occupational Safety and Health. *International Journal of Environmental Research and Public Health*, *19*(1), 349. DOI: 10.3390/ijerph19010349 PMID: 35010608

Franco, P., Cardea, S., Tabernero, A., & De Marco, I. (2021). Porous Aerogels and Adsorption of Pollutants from Water and Air: A Review. *Molecules (Basel, Switzerland), 26*(15), 4440. DOI: 10.3390/molecules26154440 PMID: 34361593

Fresco-Cala, B., Batista, A. D., & Cárdenas, S. (2020). Molecularly Imprinted Polymer Micro- and Nano-Particles. A review. *Molecules (Basel, Switzerland), 25*(20), 4740. DOI: 10.3390/molecules25204740 PMID: 33076552

Frydenberg, T., Weinell, C. E., Dam-Johansen, K., Wallström, E., & Kiil, S. (2022). Characterization and Release Mechanisms of Aerogel-Encapsulated Biocide Crystals for Low-Loading and High-Utilization Antifouling Coatings. *ACS Omega, 7*(39), 34824–34838. DOI: 10.1021/acsomega.2c03133 PMID: 36211059

Galdames, A., Ruiz-Rubio, L., Orueta, M., Sánchez-Arzalluz, M., & Vilas-Vilela, J. L. (2020). Zero-Valent Iron Nanoparticles for Soil and Groundwater Remediation. *International Journal of Environmental Research and Public Health, 17*(16), 5817. DOI: 10.3390/ijerph17165817 PMID: 32796749

Gandla, K., Kumar, K. P., Rajasulochana, P., Charde, M. S., Rana, R., Singh, L. P., Haque, M. A., Bakshi, V., Siddiqui, F. A., Khan, S. L., & Ganguly, S. (2023). Fluorescent-Nanoparticle-Impregnated Nanocomposite Polymeric Gels for Biosensing and Drug Delivery Applications. *Gels (Basel, Switzerland), 9*(8), 669. DOI: 10.3390/gels9080669 PMID: 37623124

Ganobjak, M., Brunner, S., Hofmann, J., Klar, V., Ledermann, M., Herzog, V., Kämpfen, B., Kilian, R., Wehdorn, M., & Wernery, J. (2023). Current Trends in Aerogel Use in Heritage Buildings: Case Studies from the Aerogel Architecture Award 2021. *Gels (Basel, Switzerland), 9*(10), 814. DOI: 10.3390/gels9100814 PMID: 37888387

Gao, J., Ma, R., Feng, L., Liu, Y., Jackstell, R., Jagadeesh, R. V., & Beller, M. (2021). Ambient Hydrogenation and Deuteration of Alkenes Using a Nanostructured Ni-Core-Shell Catalyst. *Angewandte Chemie International Edition, 60*(34), 18591–18598. DOI: 10.1002/anie.202105492 PMID: 34076934

Garcia-Gil, A., Biswas, S., & Holmes, J. D. (2021). A Review of Self-Seeded Germanium Nanowires: Synthesis, Growth Mechanisms and Potential Applications. *Nanomaterials (Basel, Switzerland), 11*(8), 2002. DOI: 10.3390/nano11082002 PMID: 34443831

García-González, C. A., Budtova, T., Durães, L., Erkey, C., Del Gaudio, P., Gurikov, P., Koebel, M., Liebner, F., Neagu, M., & Smirnova, I. (2019). An Opinion Paper on Aerogels for Biomedical and Environmental Applications. *Molecules (Basel, Switzerland), 24*(9), 1815. DOI: 10.3390/molecules24091815 PMID: 31083427

Gatou, M. A., Vagena, I. A., Lagopati, N., Pippa, N., Gazouli, M., & Pavlatou, E. A. (2023). Functional MOF-Based Materials for Environmental and Biomedical Applications: A Critical Review. *Nanomaterials (Basel, Switzerland), 13*(15), 2224. DOI: 10.3390/nano13152224 PMID: 37570542

Gavrila , S., Ursachi, C. ., Per a-Cri an, S., & Munteanu, F. D. (2022). Recent Trends in Biosensors for Environmental Quality Monitoring. *Sensors (Basel), 22*(4), 1513. DOI: 10.3390/s22041513 PMID: 35214408

Ghica, M. E., Mandinga, J. G. S., Linhares, T., Almeida, C. M. R., & Durães, L. (2023). Improvement of the Mechanical Properties of Silica Aerogels for Thermal Insulation Applications through a Combination of Aramid Nanofibres and Microfibres. *Gels (Basel, Switzerland), 9*(7), 535. DOI: 10.3390/gels9070535 PMID: 37504414

Glatz, R. T., Ates, H. C., Mohsenin, H., Weber, W., & Dincer, C. (2022). Designing electrochemical microfluidic multiplexed biosensors for on-site applications. *Analytical and Bioanalytical Chemistry, 414*(22), 6531–6540. DOI: 10.1007/s00216-022-04210-4 PMID: 35794347

Goyal, D., Dang, R. K., Goyal, T., Saxena, K. K., Mohammed, K. A., & Dixit, S. (2022). Graphene: A Path-Breaking Discovery for Energy Storage and Sustainability. *Materials (Basel), 15*(18), 6241. DOI: 10.3390/ma15186241 PMID: 36143552

Greszta, A., Bartkowiak, G., Dąbrowska, A., Gliścińska, E., Machnowski, W., & Kozikowski, P. (2022). Multilayer Nonwoven Inserts with Aerogel/PCMs for the Improvement of Thermophysiological Comfort in Protective Clothing against the Cold. *Materials (Basel), 15*(6), 2307. DOI: 10.3390/ma15062307 PMID: 35329760

Grisolia, A., Dell'Olio, G., Spadafora, A., De Santo, M., Morelli, C., Leggio, A., & Pasqua, L. (2023). Hybrid Polymer-Silica Nanostructured Materials for Environmental Remediation. *Molecules (Basel, Switzerland), 28*(13), 5105. DOI: 10.3390/molecules28135105 PMID: 37446768

Guastaferro, M., Baldino, L., Reverchon, E., & Cardea, S. (2021). Production of Porous Agarose-Based Structures: Freeze-Drying vs. Supercritical CO_2 Drying. *Gels (Basel, Switzerland), 7*(4), 198. DOI: 10.3390/gels7040198 PMID: 34842697

Guastaferro, M., Reverchon, E., & Baldino, L. (2021). Polysaccharide-Based Aerogel Production for Biomedical Applications: A Comparative Review. *Materials (Basel), 14*(7), 1631. DOI: 10.3390/ma14071631 PMID: 33810582

Guengerich, F. P., & Yoshimoto, F. K. (2018). Formation and Cleavage of C-C Bonds by Enzymatic Oxidation-Reduction Reactions. *Chemical Reviews, 118*(14), 6573–6655. DOI: 10.1021/acs.chemrev.8b00031 PMID: 29932643

Guerra, F. D., Attia, M. F., Whitehead, D. C., & Alexis, F. (2018). Nanotechnology for Environmental Remediation: Materials and Applications. *Molecules (Basel, Switzerland), 23*(7), 1760. DOI: 10.3390/molecules23071760 PMID: 30021974

Guzzinati, G., Altantzis, T., Batuk, M., De Backer, A., Lumbeeck, G., Samaee, V., Batuk, D., Idrissi, H., Hadermann, J., Van Aert, S., Schryvers, D., Verbeeck, J., & Bals, S. (2018). Recent Advances in Transmission Electron Microscopy for Materials Science at the EMAT Lab of the University of Antwerp. *Materials (Basel), 11*(8), 1304. DOI: 10.3390/ma11081304 PMID: 30060556

Hamachi, L. S., Yang, H., Jen-La Plante, I., Saenz, N., Qian, K., Campos, M. P., Cleveland, G. T., Rreza, I., Oza, A., Walravens, W., Chan, E. M., Hens, Z., Crowther, A. C., & Owen, J. S. (2019). Precursor reaction kinetics control compositional grading and size of CdSe1-x S x nanocrystalheterostructures. *Chemical Science (Cambridge), 10*(26), 6539–6552. DOI: 10.1039/C9SC00989B PMID: 31367306

Han, F., Li, M., Ye, H., & Zhang, G. (2021). Materials, Electrical Performance, Mechanisms, Applications, and Manufacturing Approaches for Flexible Strain Sensors. *Nanomaterials (Basel, Switzerland), 11*(5), 1220. DOI: 10.3390/nano11051220 PMID: 34063165

Harun-Ur-Rashid, M., Jahan, I., Foyez, T., & Imran, A. B. (2023). Bio-Inspired Nanomaterials for Micro/Nanodevices: A New Era in Biomedical Applications. *Micromachines, 14*(9), 1786. DOI: 10.3390/mi14091786 PMID: 37763949

Hayes, G., Laurel, M., MacKinnon, D., Zhao, T., Houck, H. A., & Becer, C. R. (2023). Polymers without Petrochemicals: Sustainable Routes to Conventional Monomers. *Chemical Reviews, 123*(5), 2609–2734. DOI: 10.1021/acs.chemrev.2c00354 PMID: 36227737

He, D., Cui, Y., Ming, F., & Wu, W. (2023). Advancements in Passive Wireless Sensors, Materials, Devices, and Applications. *Sensors (Basel), 23*(19), 8200. DOI: 10.3390/s23198200 PMID: 37837030

Horvat, G., & Maver, U. (2024). Editorial on the Special Issue Entitled "Recent Advances in Aerogels". *Gels (Basel, Switzerland), 10*(3), 154. DOI: 10.3390/gels10030154 PMID: 38534572

Horvat, G., Pantić, M., Knez, Ž., & Novak, Z. (2022). A Brief Evaluation of Pore Structure Determination for Bioaerogels. *Gels (Basel, Switzerland), 8*(7), 438. DOI: 10.3390/gels8070438 PMID: 35877523

Hussain, S., Mubeen, I., Ullah, N., Shah, S. S. U. D., Khan, B. A., Zahoor, M., Ullah, R., Khan, F. A., & Sultan, M. A. (2022). Modern Diagnostic Imaging Technique Applications and Risk Factors in the Medical Field: A Review. *BioMed Research International, 5164970*(1), 5164970. Advance online publication. DOI: 10.1155/2022/5164970 PMID: 35707373

Inobeme, A., Adetunji, C. O., Mathew, J T., Ajai, A. I., Inobeme, J., Bamigboye, M. O., Onyeaku, S., Maliki, M., Eziukwu, C. &Tawa, K. (2023). Nanotechnology for Bioremediation of Heavy Metals. In book: Microbial Technologies in Industrial Wastewater Treatment. Springer Nature Singapore, 19-30. .DOI: 10.1007/978-981-99-2435-6_2

Inobeme, A., Adetunji, C. O., Mathew, J. T., Ajai, I. A., & Jacob, J. O. (2024)b. Characterization of Plant-Based Nanobiofungicides. In book: Handbook of Agricultural Biotechnology, John Wiley & Sons, 81-97. DOI: DOI: 10.1002/9781119836544.ch5

Inobeme, A., Mathew, J.T., Adetunji, C.O., Agbugui, M.O., Osarenren, E.J., Oti, C., Ngonso, B.F., Inobeme, J., Oyedolapo, M.B., Bernard, E. &Jibrin, N.A. (2024). Biocomposites with graphene derivatives. In book: Advances in Biocomposites and their Applications, Elsevier Inc., 149-166. DOI: DOI: 10.1016/B978-0-443-19074-2.00005-8

Inobeme, A., Shahnawaz, M., Adetunji, C. O., Mathew, J. T., & Adetuyi, B. O. Popoola, Oluwabukola Atinuke; Olaitan, Falana Yetunde; Akinbo, O., Kolawole, O. M., Oyewole, O. A., Eniola, K.I.T., Yerima, M. B. (2024)c. Primary Microplastic: Source, Uses, Transportation from Land to Marine Environment. In book: Microplastic Pollution. Springer, Singapore, 1, 23–33. DOI: DOI: 10.1007/978-981-99-8357-5_2

Inobeme, A., Shahnawaz, M., Adetunji, C. O., Mathew, J. T., Adetuyi, B. O., Popoola, O. A., Olaitan, F. Y., Akinbo, O., Kolawole, O. M., Oyewole, O. A., Eniola, K. I. T., & Yerima, B. M. (2024)d. Microplastic as a Pathogenic Vector to Cause Diseases in Marine Biota. In book: Microplastic Pollution. Springer, Singapore, 1, 321–349. DOI: DOI: 10.1007/978-981-99-8357-5_19

Inobeme, A., Adetunji, C. O., Mathew, J. T., Okonkwo, S., Bamigboye, M. O., Ajai, A. I., Afoso, E., & Inobeme, J. (2023). Advanced nanotechnology for the degradation of persistent organic pollutants. In Shah, M. (Ed.), *Microbial Degradation and Detoxification of Pollutants* (pp. 51–72). De Gruyter., DOI: 10.1515/9783110743623-003

Inobeme, A., Nayak, V., Mathew, J. T., Okonkwo, S., Ekwoba, L., Ajai, A. I., Bernard, E., Inobeme, J., Agbugui, M. M., & Singh, K. R. B. (2022). Chemometric approach in environmental pollution analysis: A critical review. *Journal of Environmental Management*, *309*(114653), 1–18. DOI: 10.1016/j.jenvman.2022.114653 PMID: 35176568

Jiang, Z., Zheng, Z., Yu, S., Gao, Y., Ma, J., Huang, L., & Yang, L. (2023). Nanofiber Scaffolds as Drug Delivery Systems Promoting Wound Healing. *Pharmaceutics*, *15*(7), 1829. DOI: 10.3390/pharmaceutics15071829 PMID: 37514015

Jia, S., Gao, H., Xue, Z., & Meng, X. (2022). Recent Advances in Multifunctional Wearable Sensors and Systems: Design, Fabrication, and Applications. *Biosensors (Basel)*, *12*(11), 1057. DOI: 10.3390/bios12111057 PMID: 36421175

Jin, R., Zhou, Z., Liu, J., Shi, B., Zhou, N., Wang, X., Jia, X., Guo, D., & Xu, B. (2023). Aerogels for Thermal Protection and Their Application in Aerospace. *Gels (Basel, Switzerland)*, *9*(8), 606. DOI: 10.3390/gels9080606 PMID: 37623061

Jóźwiak, B., Dzido, G., Zor bski, E., Kolanowska, A., J drysiak, R., Dziadosz, J., Libera, M., Boncel, S., & Dzida, M. (2020). Remarkable Thermal Conductivity Enhancement in Carbon-Based Ionanofluids: Effect of Nanoparticle Morphology. *ACS Applied Materials & Interfaces, 12*(34), 38113–38123. DOI: 10.1021/acsami.0c09752 PMID: 32649171

Juhász, L., Moldován, K., Gurikov, P., Liebner, F., Fábián, I., Kalmár, J., & Cserháti, C. (2021). False Morphology of Aerogels Caused by Gold Coating for SEM Imaging. *Polymers, 13*(4), 588. DOI: 10.3390/polym13040588 PMID: 33669181

Kamanina, O. A., Saverina, E. A., Rybochkin, P. V., Arlyapov, V. A., Vereshchagin, A. N., & Ananikov, V. P. (2022). Preparation of Hybrid Sol-Gel Materials Based on Living Cells of Microorganisms and Their Application in Nanotechnology. *Nanomaterials (Basel, Switzerland), 12*(7), 1086. DOI: 10.3390/nano12071086 PMID: 35407203

Kang, X., Yip, S., Meng, Y., Wang, W., Li, D., Liu, C., & Ho, J. C. (2021). High-performance electrically transduced hazardous gas sensors based on low-dimensional nanomaterials. *Nanoscale Advances, 3*(22), 6254–6270. DOI: 10.1039/D1NA00433F PMID: 36133491

Kantaros, A., Ganetsos, T., & Petrescu, F. I. T. (2024). Transforming Object Design and Creation: Biomaterials and Contemporary Manufacturing Leading the Way. *Biomimetics (Basel, Switzerland), 9*(1), 48. DOI: 10.3390/biomimetics9010048 PMID: 38248622

Karamikamkar, S., Yalcintas, E. P., Haghniaz, R., de Barros, N. R., Mecwan, M., Nasiri, R., Davoodi, E., Nasrollahi, F., Erdem, A., Kang, H., Lee, J., Zhu, Y., Ahadian, S., Jucaud, V., Maleki, H., Dokmeci, M. R., Kim, H. J., &Khademhosseini, A. (2023). Aerogel-Based Biomaterials for Biomedical Applications: From Fabrication Methods to Disease-Targeting Applications. Advanced science (Weinheim, Baden-Wurttemberg, Germany), 10(23), e2204681. DOI: 10.1002/advs.202204681

Katrilaka, C., Karipidou, N., Petrou, N., Manglaris, C., Katrilakas, G., Tzavellas, A. N., Pitou, M., Tsiridis, E. E., Choli-Papadopoulou, T., & Aggeli, A. (2023). Freeze-Drying Process for the Fabrication of Collagen-Based Sponges as Medical Devices in Biomedical Engineering. *Materials (Basel), 16*(12), 4425. DOI: 10.3390/ma16124425 PMID: 37374608

Kausar, A., Ahmad, I., Zhao, T., Aldaghri, O., Ibnaouf, K. H., & Eisa, M. H. (2023). GrapheneNanocomposites as Innovative Materials for Energy Storage and Conversion-Design and Headways. *International Journal of Molecular Sciences, 24*(14), 11593. DOI: 10.3390/ijms241411593 PMID: 37511354

Keller, J. G., Wiemann, M., Gröters, S., Werle, K., Vennemann, A., Landsiedel, R., & Wohlleben, W. (2021). Aerogels are not regulated as nanomaterials, but can be assessed by tiered testing and grouping strategies for nanomaterials. *Nanoscale Advances, 3*(13), 3881–3893. DOI: 10.1039/D1NA00044F PMID: 36133012

Keshavarz, L., Ghaani, M. R., & English, N. J. (2021). The Importance of Precursors and Modification Groups of Aerogels in CO2 Capture. *Molecules (Basel, Switzerland), 26*(16), 5023. DOI: 10.3390/molecules26165023 PMID: 34443610

Khandelwal, K., Boahene, P., Nanda, S., & Dalai, A. K. (2023). A Review of the Design and Performance of Catalysts for Hydrothermal Gasification of Biomass to Produce Hydrogen-Rich Gas Fuel. *Molecules (Basel, Switzerland), 28*(13), 5137. DOI: 10.3390/molecules28135137 PMID: 37446799

Khan, I. U., Qi, S. S., Gul, F., Manan, S., Rono, J. K., Naz, M., Shi, X. N., Zhang, H., Dai, Z. C., & Du, D. L. (2023). A Green Approach Used for Heavy Metals 'Phytoremediation' Via Invasive Plant Species to Mitigate Environmental Pollution: A Review. *Plants, 12*(4), 725. DOI: 10.3390/plants12040725 PMID: 36840073

Khan, N. R., Sharmin, T., & Bin Rashid, A. (2023). Exploring the Versatility of Aerogels: Broad Applications in Biomedical Engineering, Astronautics, Energy Storage, Biosensing, and Current Progress. *Heliyon, 10*(1), e23102. DOI: 10.1016/j.heliyon.2023.e23102 PMID: 38163169

Kim, J.-H., Ahn, J.-H., Kim, J.-D., Lee, D.-H., Kim, S.-K., & Lee, J.-M. (2021). Influence of Silica-Aerogel on Mechanical Characteristics of Polyurethane-Based Composites: Thermal Conductivity and Strength. *Materials (Basel), 14*(7), 1790. DOI: 10.3390/ma14071790 PMID: 33916354

Kim, Y., Baek, S., Gupta, P., Kim, C., Chang, K., Ryu, S. P., Kang, H., Kim, W. S., Myoung, J., Park, W., & Kim, K. (2019). Air-like plasmonics with ultralow-refractive-index silica aerogels. *Scientific Reports, 9*(1), 2265. DOI: 10.1038/s41598-019-38859-2 PMID: 30783170

Koehle, A.P., Brumwell, S.L. &Seto, E.P. (2023). Microbial applications for sustainable space exploration beyond low Earth orbit. *npj Microgravity* 9, 47. DOI: 10.1038/s41526-023-00285-0

Kravanja, K. A., Finšgar, M., Knez, Ž., & Knez Marevci, M. (2022). Supercritical Fluid Technologies for the Incorporation of Synthetic and Natural Active Compounds into Materials for Drug Formulation and Delivery. *Pharmaceutics, 14*(8), 1670. DOI: 10.3390/pharmaceutics14081670 PMID: 36015296

Kubovics, M., Silva, C. G., López-Periago, A. M., Faria, J. L., & Domingo, C. (2022). Photocatalytic Hydrogen Production using Porous 3D Graphene-Based Aerogels Supporting Pt/TiO2 Nanoparticles. *Gels (Basel, Switzerland), 8*(11), 719. DOI: 10.3390/gels8110719 PMID: 36354627

Kumar, D., Alam, M., & Sanjayan, J. G. (2021). Retrofitting Building Envelope Using Phase Change Materials and Aerogel Render for Adaptation to Extreme Heatwave: A Multi-Objective Analysis Considering Heat Stress, Energy, Environment, and Cost. *Sustainability (Basel), 13*(19), 10716. DOI: 10.3390/su131910716

Lamy-Mendes, A., Lopes, D., Girão, A. V., Silva, R. F., Malfait, W. J., & Durães, L. (2023). Carbon Nanostructures-Silica Aerogel Composites for Adsorption of Organic Pollutants. *Toxics*, *11*(3), 232. DOI: 10.3390/toxics11030232 PMID: 36976997

Lamy-Mendes, A., Torres, R. B., Vareda, J. P., Lopes, D., Ferreira, M., Valente, V., Girão, A. V., Valente, A. J. M., & Durães, L. (2019). Amine Modification of Silica Aerogels/Xerogels for Removal of Relevant Environmental Pollutants. *Molecules (Basel, Switzerland)*, *24*(20), 3701. DOI: 10.3390/molecules24203701 PMID: 31618901

Larrañaga-Tapia, M., Betancourt-Tovar, B., Videa, M., Antunes-Ricardo, M., & Cholula-Díaz, J. L. (2023). Green synthesis trends and potential applications of bimetallic nanoparticles towards the sustainable development goals 2030. *Nanoscale Advances*, *6*(1), 51–71. DOI: 10.1039/D3NA00761H PMID: 38125589

Lau, J., Rousseau, E., Kwon, D., Lin, K. S., Bénard, F., & Chen, X. (2020). Insight into the Development of PET Radiopharmaceuticals for Oncology. *Cancers (Basel)*, *12*(5), 1312. DOI: 10.3390/cancers12051312 PMID: 32455729

Lázár, I., Čelko, L., & Menelaou, M. (2023). Aerogel-Based Materials in Bone and Cartilage Tissue Engineering-A Review with Future Implications. *Gels (Basel, Switzerland)*, *9*(9), 746. DOI: 10.3390/gels9090746 PMID: 37754427

Lázár, I., & Menelaou, M. (2023). Editorial for the Special Issue: "Aerogel Hybrids and Nanocomposites". *Gels (Basel, Switzerland)*, *9*(10), 812. DOI: 10.3390/gels9100812 PMID: 37888385

Lebedev, A., Suslova, E., Troyankin, A., & Lovskaya, D. (2021). Investigation of Aerogel Production Processes: Solvent Exchange under High Pressure Combined with Supercritical Drying in One Apparatus. *Gels (Basel, Switzerland)*, *7*(1), 4. DOI: 10.3390/gels7010004 PMID: 33466392

Le, C. V., & Yoon, H. (2024). Advances in the Use of Conducting Polymers for Healthcare Monitoring. *International Journal of Molecular Sciences*, *25*(3), 1564. DOI: 10.3390/ijms25031564 PMID: 38338846

Lee, K. H., Arshad, Z., Dahshan, A., Alshareef, M., Alsulami, Q. A., Bibi, A., Lee, E.-J., Nawaz, M., Zubair, U., & Javid, A. (2023). Porous Aerogel Structures as Promising Materials for Photocatalysis, Thermal Insulation Textiles, and Technical Applications: A Review. *Catalysts*, *13*(9), 1286. DOI: 10.3390/catal13091286

Lermontov, S. A., Vlasenko, N. E., Sipyagina, N. A., Malkova, A. N., Gozhikova, I. O., Baranchikov, A. E., & Knerelman, E. I. (2023). Highly Porous Para-Aramid Aerogel as a Heterogeneous Catalyst for Selective Hydrogenation of Unsaturated Organic Compounds. *Polymers*, *15*(15), 3206. DOI: 10.3390/polym15153206 PMID: 37571100

Leventis, N. (2022). Polyurea Aerogels: Synthesis, Material Properties, and Applications. *Polymers*, *14*(5), 969. DOI: 10.3390/polym14050969 PMID: 35267798

Liang, Y., Li, C., Liu, Z., Wang, X., Zeng, F., Yuan, X., & Pan, Y. (2023). Decarbonization potentials of the embodied energy use and operational process in buildings: A review from the life-cycle perspective. *Heliyon*, *9*(10), e20190. DOI: 10.1016/j.heliyon.2023.e20190 PMID: 37810847

Li, H., Wang, L., Wei, Y., Yan, W., & Feng, J. (2022). Preparation of Templated Materials and Their Application to Typical Pollutants in Wastewater: A Review. *Frontiers in Chemistry*, *10*, 882876. DOI: 10.3389/fchem.2022.882876 PMID: 35480393

Li, L. J., Xia, W. J., Ma, G. P., Chen, Y. L., & Ma, Y. Y. (2019). A study on the enzymatic properties and reuse of cellulaseimmobilized with carbon nanotubes and sodium alginate. *AMB Express*, *9*(1), 112. DOI: 10.1186/s13568-019-0835-0 PMID: 31332555

Li, L., Xu, T., Zhang, F., Du, C., & He, S. (2023). Preparation of Super-Flexible Silica Aerogel and Its Application in Oil-Water Separation. *Gels (Basel, Switzerland)*, *9*(9), 739. DOI: 10.3390/gels9090739 PMID: 37754420

Linhares, T., Carneiro, V. H., Pessoa de Amorim, M. T., & Durães, L. (2023). A Comparative Thermoacoustic Insulation Study of Silica Aerogels Reinforced with Reclaimed Textile Fibres: Cotton, Polyester and Wool. *Gels (Basel, Switzerland), 9*(7), 548. DOI: 10.3390/gels9070548 PMID: 37504426

Li, P., Yang, C., Xu, X., Miao, C., He, T., Jiang, B., & Wu, W. (2022). Preparation of Bio-Based Aerogel and Its Adsorption Properties for Organic Dyes. *Gels (Basel, Switzerland), 8*(11), 755. DOI: 10.3390/gels8110755 PMID: 36421576

Li, S., Li, H., Lu, Y., Zhou, M., Jiang, S., Du, X., & Guo, C. (2023). Advanced Textile-Based Wearable Biosensors for Healthcare Monitoring. *Biosensors (Basel), 13*(10), 909. DOI: 10.3390/bios13100909 PMID: 37887102

Liu, C., Wu, S., Yang, Z., Sun, H., Zhu, Z., Liang, W., & Li, A. (2020). Mechanically Robust and Flame-Retardant Silicon Aerogel Elastomers for Thermal Insulation and Efficient Solar Steam Generation. *ACS Omega, 5*(15), 8638–8646. DOI: 10.1021/acsomega.0c00086 PMID: 32337427

Liu, J., Buahom, P., Lu, C., Yu, H., & Park, C. B. (2022). Microscopic revelation of the solid-gas coupling and Knudsen effect on the thermal conductivity of silica aerogel with inter-connected pores. *Scientific Reports, 12*(1), 21034. DOI: 10.1038/s41598-022-24133-5 PMID: 36470925

Liu, J., Yang, L., Liu, K., & Gao, F. (2023). Hydrogel scaffolds in bone regeneration: Their promising roles in angiogenesis. *Frontiers in Pharmacology, 14*, 1050954. DOI: 10.3389/fphar.2023.1050954 PMID: 36860296

Liu, Z., Xu, B., Jiang, Y. J., Zhou, Y., Sun, X., Wang, Y., & Zhu, W. (2023). Photocatalytic Conversion of Methane: Current State of the Art, Challenges, and Future Perspectives. *ACS Environmental Au, 3*(5), 252–276. DOI: 10.1021/acsenvironau.3c00002 PMID: 37743954

Li, W., & Liu, Z. (2023). Social, Environmental, and Governance Factors on Supply-Chain Performance with Mediating Technology Adoption. *Sustainability (Basel), 15*(14), 10865. DOI: 10.3390/su151410865

Long, L. Y., Weng, Y. X., & Wang, Y. Z. (2018). Cellulose Aerogels: Synthesis, Applications, and Prospects. *Polymers*, *10*(6), 623. DOI: 10.3390/polym10060623 PMID: 30966656

Lou, F., Dong, S., Zhu, K., Chen, X., & Ma, Y. (2023). Thermal Insulation Performance of Aerogel Nano-Porous Materials: Characterization and Test Methods. *Gels (Basel, Switzerland)*, *9*(3), 220. DOI: 10.3390/gels9030220 PMID: 36975669

Luengo Morato, Y., Ovejero Paredes, K., Lozano Chamizo, L., Marciello, M., & Filice, M.LuengoMorato. (2021). Recent Advances in Multimodal Molecular Imaging of Cancer Mediated by Hybrid Magnetic Nanoparticles. *Polymers*, *13*(17), 2989. DOI: 10.3390/polym13172989 PMID: 34503029

Luo, Y., Mei, Y., Xu, Y., & Huang, K. (2023). Hyper-Crosslinked Porous Organic Nanomaterials: Structure-Oriented Design and Catalytic Applications. *Nanomaterials (Basel, Switzerland)*, *13*(18), 2514. DOI: 10.3390/nano13182514 PMID: 37764543

Ma, B., Cheng, Y., Hu, P., Fang, D., & Wang, J. (2023). Passive Daytime Radiative Cooling of Silica Aerogels. *Nanomaterials (Basel, Switzerland)*, *13*(3), 467. DOI: 10.3390/nano13030467 PMID: 36770428

Mahmoud, L. A. M., Dos Reis, R. A., Chen, X., Ting, V. P., & Nayak, S. (2022). Metal-Organic Frameworks as Potential Agents for Extraction and Delivery of Pesticides and Agrochemicals. *ACS Omega*, *7*(50), 45910–45934. DOI: 10.1021/acsomega.2c05978 PMID: 36570238

Maio, A., Pibiri, I., Morreale, M., Mantia, F. P., & Scaffaro, R. (2021). An Overview of Functionalized GrapheneNanomaterials for Advanced Applications. *Nanomaterials (Basel, Switzerland)*, *11*(7), 1717. DOI: 10.3390/nano11071717 PMID: 34209928

Majerič, P., & Rudolf, R. (2020). Advances in Ultrasonic Spray Pyrolysis Processing of Noble Metal Nanoparticles-Review. *Materials (Basel)*, *13*(16), 3485. DOI: 10.3390/ma13163485 PMID: 32784637

Mallakpour, S., Azadi, E., & Hussain, C. M. (2022). Fabrication of air filters with advanced filtration performance for removal of viral aerosols and control the spread of COVID-19. *Advances in Colloid and Interface Science, 303*, 102653. DOI: 10.1016/j.cis.2022.102653 PMID: 35349924

Maponya, T. C., Makgopa, K., Somo, T. R., & Modibane, K. D. (2022). Highlighting the Importance of Characterization Techniques Employed in Adsorption Using Metal-Organic Frameworks for Water Treatment. *Polymers, 14*(17), 3613. DOI: 10.3390/polym14173613 PMID: 36080689

Mathew, J. T., Adetunji, C. O., Inobeme, A., Musah, M., Azeh, Y., Otori, A. A., Shaba, E. Y., Mamman, A., & Tanko, E. (2023)b. Removal of Heavy Metals Using Bio-remedial Techniques. Springer Nature Switzerland AG 2023 M. P. Shah (ed.), Modern Approaches in Waste Bioremediation, 117-130. DOI: 10.1007/978-3-031-24086-7_6

Mathew, J. T., Inobeme, A., Adetuyi, B. O., Adetunji, C. O., Popoola, O. A., Olaitan, F. Y., Akinbo, O., Shahnawaz, M., Oyewole, O. A., Eniola, K. I. T., & Yerima, B. M. (2024)d. General Introduction of Microplastic: Uses, Types, and Generation. In book: Microplastic Pollution. Springer, Singapore, 1, 3-21. DOI: DOI: 10.1007/978-981-99-8357-5_1

Mathew, J. T., Inobeme, A., Etsuyankpa, M. B., Adetunji, C. O., Shaba, E. Y., Musah, M., Azeh, Y., Jiya, M. J., Tanko, E., Muhammad, A. I., Mamman, A. & Maurice, J. (2024)d. Applications of Diatom for Treatment of Industrial Effluents: Waste to Wealth Approach.. In book: Marine Biomass: Biorefinery, Bioproducts and Environmental Bioremediation. Walter de Gruyter GmbH & Co KG, 1-20, DOI: DOI: 10.1515/9783111353951-001

Mathew, J. T., Inobeme, A., Abdulkadir, A., Etsuyankpa, M. B., Shaba, E. Y., Musah, M., Azeh, Y., Adetunji, C. O., Ibrahim, A. M., Tanko, E., & Mamman, A. (2024). a. Properties and Characteristics of Nanocoatings for Medicinal Applications. In book: Sustainable Approach to Protective Nanocoatings. IGI Global; publishing Tomorrow's. *Advances in Chemical and Materials Engineering*, •••, 203–227. DOI: 10.4018/979-8-3693-3136-1.ch007

Mathew, J. T., Inobeme, A., Azeh, Y., Musah, M., Abdulkadir, A., Shaba, E. Y., Etsuyankpa, M. B., Musa, T. S., Muhammad, A. I., Ismail, H., Kanwa, A. M., Mamman, A., & Hussaini, J. (2024). c. Carbon Capture and Storage Via Electrochemical and Bioelectrochemical Techniques: A Review. *Caliphate Journal of Science & Technology*, *2*(2), 141–158. DOI: 10.4314/cajost.v6i2.3

Mathew, J. T., Inobeme, A., Musah, M., Azeh, Y., Abdullahi, A., Shaba, E. Y., Salihu, A. M., Muhammad, E. B., Josiah, J. G., Jibrin, N. A., Ismail, H., Muhammad, A. I., Maurice, J., Mamman, A., & Ndamitso, M. M. (2024). b. A Critical Review of Green Approach on Wastewater Treatment Strategies. *Journal of Applied Science & Environmental Management*, *28*(2), 363–391. DOI: 10.4314/jasem.v28i2.9

Mathew, J. T., Musah, M., Azeh, Y., & Muhammed, M. (2023). a. Kinetic Study of Heavy Metals Removal from Pharmaceutical Wastewater Using Geopolymer/Fe_3O_4Nanocomposite. *Bima Journal of Science and Technology*, *7*(4), 152–163. DOI: 10.56892/bima.v7i4.539

Mathew, J. T., Musah, M., Azeh, Y., & Muhammed, M. (2023). b. Adsorptive Removal of Selected Toxic Metals from Pharmaceutical Wastewater using Fe_3O_4/ZnONanocomposite. *Dutse Journal of Pure and Applied Sciences*, *9*(4a), 236–248. DOI: 10.4314/dujopas.v9i4a.22

Mathew, J. T., Musah, M., Azeh, Y., & Muhammed, M. (2024). b. Development of Fe3O4 Nanoparticles for the Removal of Some Toxic Metals from Pharmaceutical Wastewater. [CaJoST]. *Caliphate Journal of Science & Technology*, *6*(1), 26–34. DOI: 10.4314/cajost.v6i1.4

Mazzà, C., Alcock, L., Aminian, K., Becker, C., Bertuletti, S., Bonci, T., Brown, P., Brozgol, M., Buckley, E., Carsin, A. E., Caruso, M., Caulfield, B., Cereatti, A., Chiari, L., Chynkiamis, N., Ciravegna, F., Del Din, S., Eskofier, B., Evers, J., & Rochester, L. (2021). Technical validation of real-world monitoring of gait: A multicentric observational study. *BMJ Open*, *11*(12), e050785. DOI: 10.1136/bmjopen-2021-050785 PMID: 34857567

McKay, A. I., Bukvic, A. J., Tegner, B. E., Burnage, A. L., Martı Nez-Martı Nez, A. J., Rees, N. H., Macgregor, S. A., & Weller, A. S. (2019). Room Temperature Acceptorless Alkane Dehydrogenation from Molecular σ-Alkane Complexes. *Journal of the American Chemical Society, 141*(29), 11700–11712. DOI: 10.1021/jacs.9b05577 PMID: 31246012

Meti, P., Wang, Q., Mahadik, D. B., Lee, K. Y., Gong, Y. D., & Park, H. H. (2023). Evolutionary Progress of Silica Aerogels and Their Classification Based on Composition: An Overview. *Nanomaterials (Basel, Switzerland), 13*(9), 1498. DOI: 10.3390/nano13091498 PMID: 37177045

Mosińska, M., Szynkowska-Jóźwik, M. I., & Mierczyński, P. (2020). Catalysts for Hydrogen Generation via Oxy-Steam Reforming of Methanol Process. *Materials (Basel), 13*(24), 5601. DOI: 10.3390/ma13245601 PMID: 33302526

Nallusamy, J., & Das, R. K. (2021). Hydrogels and Their Role in Bone Tissue Engineering: An Overview. *Journal of Pharmacy & Bioallied Sciences, 13*(Suppl 2), S908–S912. DOI: 10.4103/jpbs.jpbs_237_21 PMID: 35017896

Nan, Z., Wei, W., Lin, Z., Chang, J., & Hao, Y. (2023). Flexible Nanocomposite Conductors for Electromagnetic Interference Shielding. *Nano-Micro Letters, 15*(1), 172. DOI: 10.1007/s40820-023-01122-5 PMID: 37420119

Naresh, V., & Lee, N. (2021). A Review on Biosensors and Recent Development of Nanostructured Materials-Enabled Biosensors. *Sensors (Basel), 21*(4), 1109. DOI: 10.3390/s21041109 PMID: 33562639

Navas, D., Fuentes, S., Castro-Alvarez, A., & Chavez-Angel, E. (2021). Review on Sol-Gel Synthesis of Perovskite and Oxide Nanomaterials. *Gels (Basel, Switzerland), 7*(4), 275. DOI: 10.3390/gels7040275 PMID: 34940335

Neagu, M., Grisi, F., Pulvirenti, A., Simón-Vázquez, R., García-González, C. A., & Boccia, A. C. (2023). Updated Aspects of Safety Regulations for Biomedical Applications of Aerogel Compounds—Compendia-Like Evaluation. *Safety (Basel, Switzerland), 9*(4), 80. DOI: 10.3390/safety9040080

Nghiem, T. L., Coban, D., Tjaberings, S., & Gröschel, A. H. (2020). Recent Advances in the Synthesis and Application of Polymer Compartments for Catalysis. *Polymers, 12*(10), 2190. DOI: 10.3390/polym12102190 PMID: 32987965

Nguyen, P. T. T., Do, N. H. N., Goh, X. Y., Goh, C. J., Ong, R. H., Le, P. K., Phan-Thien, N., & Duong, H. M. (2022). Recent Progresses in Eco-Friendly Fabrication and Applications of Sustainable Aerogels from Various Waste Materials. *Waste and Biomass Valorization, 13*(4), 1825–1847. DOI: 10.1007/s12649-021-01627-3 PMID: 34745389

Niculescu, A. G., Tudorache, D. I., Bocioagă, M., Mihaiescu, D. E., Hadibarata, T., & Grumezescu, A. M. (2024). An Updated Overview of Silica Aerogel-Based Nanomaterials. *Nanomaterials (Basel, Switzerland), 14*(5), 469. DOI: 10.3390/nano14050469 PMID: 38470798

Nita, L. E., Ghilan, A., Rusu, A. G., Neamtu, I., & Chiriac, A. P. (2020). New Trends in Bio-Based Aerogels. *Pharmaceutics, 12*(5), 449. DOI: 10.3390/pharmaceutics12050449 PMID: 32414217

Niu, Y., Chen, L., & Wu, T. (2023). Recent Advances in Bioengineering Bone Revascularization Based on Composite Materials Comprising Hydroxyapatite. *International Journal of Molecular Sciences, 24*(15), 12492. DOI: 10.3390/ijms241512492 PMID: 37569875

Noman, M. T., Amor, N., Ali, A., Petrik, S., Coufal, R., Adach, K., & Fijalkowski, M. (2021). Aerogels for Biomedical, Energy and Sensing Applications. *Gels (Basel, Switzerland), 7*(4), 264. DOI: 10.3390/gels7040264 PMID: 34940324

Norizan, M. N., Shazleen, S. S., Alias, A. H., Sabaruddin, F. A., Asyraf, M. R. M., Zainudin, E. S., Abdullah, N., Samsudin, M. S., Kamarudin, S. H., & Norrrahim, M. N. F. (2022). Nanocellulose-Based Nanocomposites for Sustainable Applications: A Review. *Nanomaterials (Basel, Switzerland), 12*(19), 3483. DOI: 10.3390/nano12193483 PMID: 36234612

Nowak, D., & Jakubczyk, E. (2020). The Freeze-Drying of Foods-The Characteristic of the Process Course and the Effect of Its Parameters on the Physical Properties of Food Materials. *Foods*, *9*(10), 1488. DOI: 10.3390/foods9101488 PMID: 33080983

Obeid, E., & Younes, K. (2024). Uncovering Key Factors in Graphene Aerogel-Based Electrocatalysts for Sustainable Hydrogen Production: An Unsupervised Machine Learning Approach. *Gels (Basel, Switzerland)*, *10*(1), 57. DOI: 10.3390/gels10010057 PMID: 38247780

Ogbezode, J. E., Ezealigo, U. S., Bello, A., Anye, V. C., & Onwualu, A. P. (2023). A narrative review of the synthesis, characterization, and applications of iron oxide nanoparticles. *Discover Nano*, *18*(1), 125. DOI: 10.1186/s11671-023-03898-2 PMID: 37815643

Okonta, D. E. (2023). Investigating the impact of building materials on energy efficiency and indoor cooling in Nigerian homes. *Heliyon*, *9*(9), e20316. DOI: 10.1016/j.heliyon.2023.e20316 PMID: 37809702

Olorunsola, A., Adetunji, C. O., Olaniyan, O., Adetunji, J. B., Akinbo, K., Adetuyi, B. O., Inobeme, A., Mathew, J. T., & Ajenifujah-Solebo, S. O. (2024). Application of Nanotechnology for the Production of Biopesticides, Bioinsecticides, Bioherbicides, Mosquitoe Repellants and Biofungicides. In book: Handbook of Agricultural Biotechnology, John Wiley & Sons, 293-309. DOI: DOI: 10.1002/9781119836483.ch11

Owais, M., Shiverskii, A., Pal, A. K., Mahato, B., & Abaimov, S. G. (2022). Recent Studies on Thermally Conductive 3D Aerogels/Foams with the Segregated Nanofiller Framework. *Polymers*, *14*(22), 4796. DOI: 10.3390/polym14224796 PMID: 36432922

Parapouli, M., Vasileiadis, A., Afendra, A. S., & Hatziloukas, E. (2020). Saccharomyces cerevisiae and its industrial applications. *AIMS Microbiology*, *6*(1), 1–31. DOI: 10.3934/microbiol.2020001 PMID: 32226912

Paraskevopoulou, P., Chriti, D., Raptopoulos, G., & Anyfantis, G. C. (2019). Synthetic Polymer Aerogels in Particulate Form. *Materials (Basel)*, *12*(9), 1543. DOI: 10.3390/ma12091543 PMID: 31083421

Paras, Y., Yadav, K., Kumar, P., Teja, D. R., Chakraborty, S., Chakraborty, M., Mohapatra, S. S., Sahoo, A., Chou, M. M. C., Liang, C.-T., & Hang, D.-R. (2022). A Review on Low-Dimensional Nanomaterials: Nanofabrication, Characterization and Applications. *Nanomaterials (Basel, Switzerland)*, *13*(1), 160. DOI: 10.3390/nano13010160 PMID: 36616070

Payanda, K. O., Alsuhile, A. A. A. M., Yousefzadeh, H., Ulker, Z., Bozbag, S. E., García-González, C. A., Smirnova, I., & Erkey, C. (2023). The effect of synthesis conditions and process parameters on aerogel properties. *Frontiers in Chemistry*, *11*, 1294520. DOI: 10.3389/fchem.2023.1294520 PMID: 37937209

Potter, M. E., Le Brocq, J. J. M., Oakley, A. E., McShane, E. B., Vandegehuchte, B. D., & Raja, R. (2020). Butane Isomerization as a Diagnostic Tool in the Rational Design of Solid Acid Catalysts. *Catalysts*, *10*(9), 1099. DOI: 10.3390/catal10091099

Pradhan, S., Brooks, A. K., & Yadavalli, V. K. (2020). Nature-derived materials for the fabrication of functional biodevices. *Materials Today. Bio*, *7*, 100065. DOI: 10.1016/j.mtbio.2020.100065 PMID: 32613186

Priya, A. K., Muruganandam, M., Ali, S. S., & Kornaros, M. (2023). Clean-Up of Heavy Metals from Contaminated Soil by Phytoremediation: A Multidisciplinary and Eco-Friendly Approach. *Toxics*, *11*(5), 422. DOI: 10.3390/toxics11050422 PMID: 37235237

Qi, P., Zhu, H., Borodich, F., & Peng, Q. (2023). A Review of the Mechanical Properties of Graphene Aerogel Materials: Experimental Measurements and Computer Simulations. *Materials (Basel)*, *16*(5), 1800. DOI: 10.3390/ma16051800 PMID: 36902915

Ragab, A., Zouli, N., Abutaleb, A., Maafa, I. M., Ahmed, M. M., & Yousef, A. (2023). Environmental and Economic Benefits of Using Pomegranate Peel Waste for Insulation Bricks. *Materials (Basel)*, *16*(15), 5372. DOI: 10.3390/ma16155372 PMID: 37570075

Rahman, M. (2023). Magnetic Resonance Imaging and Iron-oxide Nanoparticles in the era of Personalized Medicine. *Nanotheranostics*, *7*(4), 424–449. DOI: 10.7150/ntno.86467 PMID: 37650011

Rando, G., Sfameni, S., Galletta, M., Drommi, D., Cappello, S., & Plutino, M. R. (2022). Functional Nanohybrids and Nanocomposites Development for the Removal of Environmental Pollutants and Bioremediation. *Molecules (Basel, Switzerland)*, *27*(15), 4856. DOI: 10.3390/molecules27154856 PMID: 35956804

Rapisarda, M., & Meo, M. (2022). Multifunctional Thermal, Acoustic, and Piezoresistive Properties of In Situ-Modified Composite Aerogels with Graphene Oxide as the Main Phase. *ACS Applied Materials & Interfaces*, *14*(38), 43646–43655. DOI: 10.1021/acsami.2c08042 PMID: 36121008

Rasmi, Y., Li, X., Khan, J., Ozer, T., & Choi, J. R. (2021). Emerging point-of-care biosensors for rapid diagnosis of COVID-19: Current progress, challenges, and future prospects. *Analytical and Bioanalytical Chemistry*, *413*(16), 4137–4159. DOI: 10.1007/s00216-021-03377-6 PMID: 34008124

Rauchman, S. H., Locke, B., Albert, J., De Leon, J., Peltier, M. R., & Reiss, A. B. (2023). Toxic External Exposure Leading to Ocular Surface Injury. *Vision (Basel)*, *7*(2), 32. DOI: 10.3390/vision7020032 PMID: 37092465

Revin, V. V., Nazarova, N. B., Tsareva, E. E., Liyaskina, E. V., Revin, V. D., & Pestov, N. A. (2020). Production of Bacterial Cellulose Aerogels With Improved Physico-Mechanical Properties and Antibacterial Effect. *Frontiers in Bioengineering and Biotechnology*, *8*, 603407. DOI: 10.3389/fbioe.2020.603407 PMID: 33344435

Romero-Montero, A., Valencia-Bermúdez, J. L., Rosas-Meléndez, S. A., Núñez-Tapia, I., Piña-Barba, M. C., Leyva-Gómez, G., & Del Prado-Audelo, M. L. (2023). Biopolymeric Fibrous Aerogels: The Sustainable Alternative for Water Remediation. *Polymers*, *15*(2), 262. DOI: 10.3390/polym15020262 PMID: 36679143

Rout, B. K., & Sikdar, B. K. (2017). Hazard Identification, Risk Assessment, and Control Measures as an Effective Tool of Occupational Health Assessment of Hazardous Process in an Iron Ore Pelletizing Industry. *Indian Journal of Occupational and Environmental Medicine*, *21*(2), 56–76. DOI: 10.4103/ijoem.IJOEM_19_16 PMID: 29540967

Şahin, İ., Özbakır, Y., İnönü, Z., Ulker, Z., & Erkey, C. (2017). Kinetics of Supercritical Drying of Gels. *Gels (Basel, Switzerland)*, *4*(1), 3. DOI: 10.3390/gels4010003 PMID: 30674780

Saleh, W.M., Ahmad, M.I., & Yahya, E.B., & H.P.S., A.K. (. (2023). Nanostructured Bioaerogels as a Potential Solution for Particulate Matter Pollution. *Gels (Basel, Switzerland)*, *9*, 575. DOI: 10.3390/gels9070575 PMID: 37504454

Sambucci, M., Savoni, F., & Valente, M. (2023). Aerogel Technology for Thermal Insulation of Cryogenic Tanks-Numerical Analysis for Comparison with Traditional Insulating Materials. *Gels (Basel, Switzerland)*, *9*(4), 307. DOI: 10.3390/gels9040307 PMID: 37102919

Sandoval, B. A., & Hyster, T. K. (2020). Emerging strategies for expanding the toolbox of enzymes in biocatalysis. *Current Opinion in Chemical Biology*, *55*, 45–51. DOI: 10.1016/j.cbpa.2019.12.006 PMID: 31935627

Sathish, S., & Shen, A. Q. (2021). Toward the Development of Rapid, Specific, and Sensitive Microfluidic Sensors: A Comprehensive Device Blueprint. *JACS Au*, *1*(11), 1815–1833. DOI: 10.1021/jacsau.1c00318 PMID: 34841402

Sayam, A., Rahman, A. N. M. M., Rahman, M. S., Smriti, S. A., Ahmed, F., Rabbi, M. F., Hossain, M., & Faruque, M. O. (2022). A review on carbon fiber-reinforced hierarchical composites: Mechanical performance, manufacturing process, structural applications and allied challenges. *Carbon Letters*, *32*(5), 1173–1205. DOI: 10.1007/s42823-022-00358-2

Scott, M. J., Verhagen, W. J. C., Bieber, M. T., & Marzocca, P. (2022). A Systematic Literature Review of Predictive Maintenance for Defence Fixed-Wing Aircraft Sustainment and Operations. *Sensors (Basel)*, *22*(18), 7070. DOI: 10.3390/s22187070 PMID: 36146419

Seesaard, T., & Wongchoosuk, C. (2023). Flexible and Stretchable Pressure Sensors: From Basic Principles to State-of-the-Art Applications. *Micromachines*, *14*(8), 1638. DOI: 10.3390/mi14081638 PMID: 37630177

Sena-Torralba, A., Álvarez-Diduk, R., Parolo, C., Piper, A., & Merkoçi, A. (2022). Toward Next Generation Lateral Flow Assays: Integration of Nanomaterials. *Chemical Reviews*, *122*(18), 14881–14910. DOI: 10.1021/acs.chemrev.1c01012 PMID: 36067039

Shah, K. W., & Huseien, G. F. (2020). Biomimetic Self-Healing Cementitious Construction Materials for Smart Buildings. *Biomimetics (Basel, Switzerland)*, *5*(4), 47. DOI: 10.3390/biomimetics5040047 PMID: 33049925

Shah, N., Rehan, T., Li, X., Tetik, H., Yang, G., Zhao, K., & Lin, D. (2021). Magnetic aerogel: An advanced material of high importance. *RSC Advances*, *11*(13), 7187–7204. DOI: 10.1039/D0RA10275J PMID: 35423256

Shan, J., Shan, Y., Zou, C., Hong, Y., Liu, J., & Guo, X. (2024). Cost-Effective Preparation of Hydrophobic and Thermal-Insulating Silica Aerogels. *Nanomaterials (Basel, Switzerland)*, *14*(1), 119. DOI: 10.3390/nano14010119 PMID: 38202574

Shao, Z., Chen, H., Wang, Q., Kang, G., Wang, X., Li, W., Liu, Y., & Zheng, G. (2022). High-performance multifunctional electrospun fibrous air filter for personal protection: A review. *Separation and Purification Technology*, *302*, 122175. DOI: 10.1016/j.seppur.2022.122175 PMID: 36168392

Sharma, S. K., Ranjani, P., Mamane, H., & Kumar, R. (2023). Preparation of graphene oxide-doped silica aerogel using supercritical method for efficient removal of emerging pollutants from wastewater. *Scientific Reports*, *13*(1), 16448. DOI: 10.1038/s41598-023-43613-w PMID: 37777623

Sheng, Z., Liu, Z., Hou, Y., Jiang, H., Li, Y., Li, G., & Zhang, X. (2023). The Rising Aerogel Fibers: Status, Challenges, and Opportunities. Advanced science (Weinheim, Baden-Wurttemberg, Germany), 10(9), e2205762. DOI: 10.1002/advs.202205762

Sheng, Z., Liu, Z., Hou, Y., Jiang, H., Li, Y., Li, G., & Zhang, X. (2023). *The Rising Aerogel Fibers: Status*. Challenges, and Oppo.

Shirsat, M. D., & Hianik, T. (2023). Electrochemical Detection of Heavy Metal Ions Based on Nanocomposite Materials. *Journal of Composites Science*, *7*(11), 473. DOI: 10.3390/jcs7110473

Silva, L. F. M. D., Almeida, A. G. A., Pascoal, L. M., Santos Neto, M., Lima, F. E. T., & Santos, F. S. (2022). Skin injuries due to Personal Protective Equipment and preventive measures in the COVID-19 context: an integrative review. Lesões de peleporEquipamentos de Proteção Individual e medidaspreventivas no contexto da COVID-19: revisãointegrativa. *Revista Latino-Americana de Enfermagem*, *30*, e3551. DOI: 10.1590/1518-8345.5636.3551 PMID: 35476010

Simunovic, F., & Finkenzeller, G. (2021). Vascularization Strategies in Bone Tissue Engineering. *Cells*, *10*(7), 1749. DOI: 10.3390/cells10071749 PMID: 34359919

Singh, A., Ahmed, A., Sharma, A., & Arya, S. (2022). Graphene and Its Derivatives: Synthesis and Application in the Electrochemical Detection of Analytes in Sweat. *Biosensors (Basel)*, *12*(10), 910. DOI: 10.3390/bios12100910 PMID: 36291046

Singh, G., Bhadange, S., Bhawna, F., Shewale, P., Dahiya, R., Aggarwal, A., Manju, F., & Arya, S. K. (2023). Phytoremediation of radioactive elements, possibilities and challenges: Special focus on agricultural aspects. *International Journal of Phytoremediation*, *25*(1), 1–8. DOI: 10.1080/15226514.2022.2043239 PMID: 35244498

Slepičková Kasálková, N., Slepička, P., & Švorčík, V.SlepičkováKasálková. (2021). Carbon Nanostructures, Nanolayers, and Their Composites. *Nanomaterials (Basel, Switzerland)*, *11*(9), 2368. DOI: 10.3390/nano11092368 PMID: 34578684

Smith, S., Goodge, K., Delaney, M., Struzyk, A., Tansey, N., & Frey, M. (2020). A Comprehensive Review of the Covalent Immobilization of Biomolecules onto ElectrospunNanofibers. *Nanomaterials (Basel, Switzerland)*, *10*(11), 2142. DOI: 10.3390/nano10112142 PMID: 33121181

Song, J., Choi, S., Lim, J., & Kim, K. T. (2022). Templated synthesis of microparticles with carbonaceous skeletal structures using polymer cubosomes as templates. *RSC Advances*, *12*(14), 8429–8434. DOI: 10.1039/D2RA00832G PMID: 35424782

Sozcu, S., Venkataraman, M., Wiener, J., Tomkova, B., Militky, J., & Mahmood, A. (2023). Incorporation of Cellulose-Based Aerogels into Textile Structures. *Materials (Basel)*, *17*(1), 27. DOI: 10.3390/ma17010027 PMID: 38203881

Speranza, G. (2021). Carbon Nanomaterials: Synthesis, Functionalization and Sensing Applications. *Nanomaterials (Basel, Switzerland)*, *11*(4), 967. DOI: 10.3390/nano11040967 PMID: 33918769

Sun, B., Wang, B., & Wang, R. (2023). Advances in Emerging Catalytic Materials for the Conversion of Carbon Dioxide. *Materials (Basel)*, *16*(23), 7309. DOI: 10.3390/ma16237309 PMID: 38068053

Suryanarayana, C. (2019). Mechanical Alloying: A Novel Technique to Synthesize Advanced Materials. *Research (Washington, D.C.)*, *4219812*, 2019/4219812. Advance online publication. DOI: 10.34133/2019/4219812 PMID: 31549063

Tarazona, N. A., Machatschek, R., Balcucho, J., Castro-Mayorga, J. L., Saldarriaga, J. F., & Lendlein, A. (2022). Opportunities and challenges for integrating the development of sustainable polymer materials within an international circular (bio)economy concept. *MRS Energy & Sustainability : a Review Journal*, *9*(1), 28–34. DOI: 10.1557/s43581-021-00015-7 PMID: 37521367

Tawiah, B., Ofori, E. A., & Bin, F. (2023). Scientometric Review of Sustainable Fire-Resistant Polysaccharide-Based Composite Aerogels. *Sustainability (Basel)*, *15*(16), 12185. DOI: 10.3390/su151612185

Tebes-Stevens, C., Patel, J. M., Jones, W. J., & Weber, E. J. (2017). Prediction of Hydrolysis Products of Organic Chemicals under Environmental pH Conditions. *Environmental Science & Technology*, *51*(9), 5008–5016. DOI: 10.1021/acs.est.6b05412 PMID: 28430419

Thakur, K., Attri, C., & Seth, A. (2021). Nanocarriers-based immobilization of enzymes for industrial application. 3 Biotech, 11(10), 427. DOI: 10.1007/s13205-021-02953-y

Thang, N. H., Chien, T. B., & Cuong, D. X. (2023). Polymer-Based Hydrogels Applied in Drug Delivery: An Overview. *Gels (Basel, Switzerland)*, *9*(7), 523. DOI: 10.3390/gels9070523 PMID: 37504402

Thie, C., Quallen, S., Ibrahim, A., Xing, T., & Johnson, B. (2023). Study of Energy Saving Using Silica Aerogel Insulation in a Residential Building. *Gels (Basel, Switzerland)*, *9*(2), 86. DOI: 10.3390/gels9020086 PMID: 36826255

Tinoco Navarro, L. K., & Jaroslav, C. (2023). Enhancing Photocatalytic Properties of TiO2 Photocatalyst and Heterojunctions: A Comprehensive Review of the Impact of Biphasic Systems in Aerogels and Xerogels Synthesis, Methods, and Mechanisms for Environmental Applications. *Gels (Basel, Switzerland)*, *9*(12), 976. DOI: 10.3390/gels9120976 PMID: 38131962

Tofanica, B. M., Belosinschi, D., & Volf, I. (2022). Gels, Aerogels and Hydrogels: A Challenge for the Cellulose-Based Product Industries. *Gels (Basel, Switzerland)*, *8*(8), 497. DOI: 10.3390/gels8080497 PMID: 36005098

Trache, D., Tarchoun, A. F., Derradji, M., Hamidon, T. S., Masruchin, N., Brosse, N., & Hussin, M. H. (2020). Nanocellulose: From Fundamentals to Advanced Applications. *Frontiers in Chemistry*, *8*, 392. DOI: 10.3389/fchem.2020.00392 PMID: 32435633

Uddin, M. A., Afroj, S., Hasan, T., Carr, C., Novoselov, K. S., & Karim, N. (2022). Environmental Impacts of Personal Protective Clothing Used to Combat COVID- 19. *Advanced Sustainable Systems*, *6*(1), 2100176. DOI: 10.1002/adsu.202100176 PMID: 34901387

Ullo, S. L., & Sinha, G. R. (2020). Advances in Smart Environment Monitoring Systems Using IoT and Sensors. *Sensors (Basel)*, *20*(11), 3113. DOI: 10.3390/s20113113 PMID: 32486411

Usala, E., Espinosa, E., El Arfaoui, W., Morcillo-Martín, R., Ferrari, B., & González, Z. (2023). Antibacterial Aerogels-Based Membranes by Customized Colloidal Functionalization of TEMPO-Oxidized Cellulose Nanofibers Incorporating CuO. *Bioengineering (Basel, Switzerland), 10*(11), 1312. DOI: 10.3390/bioengineering10111312 PMID: 38002436

Vagtholm, R., Matteo, A., Vand, B., & Tupenaite, L. (2023). Evolution and Current State of Building Materials, Construction Methods, and Building Regulations in the U.K.: Implications for Sustainable Building Practices. *Buildings (Basel, Switzerland), 13*(6), 1480. DOI: 10.3390/buildings13061480

Vrabič-Brodnjak, U. (2023). Hybrid Materials of Bio-Based Aerogels for Sustainable Packaging Solutions. *Gels (Basel, Switzerland), 10*(1), 27. DOI: 10.3390/gels10010027 PMID: 38247750

Wang, H., Wang, L., & Xiao, F. S. (2020). Metal@Zeolite Hybrid Materials for Catalysis. *ACS Central Science, 6*(10), 1685–1697. DOI: 10.1021/acscentsci.0c01130 PMID: 33145408

Wang, J., Petit, D., & Ren, S. (2020). Transparent thermal insulation silica aerogels. *Nanoscale Advances, 2*(12), 5504–5515. DOI: 10.1039/D0NA00655F PMID: 36133881

Wang, M., Zhu, P., Liu, S., Chen, Y., Liang, D., Liu, Y., Chen, W., Du, L., & Wu, C. (2023). Application of Nanozymes in Environmental Monitoring, Management, and Protection. *Biosensors (Basel), 13*(3), 314. DOI: 10.3390/bios13030314 PMID: 36979526

Wang, X., Liu, T., Li, H., Han, C., Su, P., Ta, N., Jiang, S. P., Kong, B., Liu, J., & Huang, Z. (2023). Balancing Mass Transfer and Active Sites to Improve Electrocatalytic Oxygen Reduction by B,NCodoped C Nanoreactors. *Nano Letters, 23*(11), 4699–4707. DOI: 10.1021/acs.nanolett.3c00202 PMID: 36951377

Wang, Z., Liu, L., Zhang, Y., Huang, Y., Liu, J., Zhang, X., Liu, X., Teng, H., Zhang, X., Zhang, J., & Yang, H. (2023). A Review of Graphene-Based Materials/Polymer Composite Aerogels. *Polymers, 15*(8), 1888. DOI: 10.3390/polym15081888 PMID: 37112034

Wu, H., Shang, D., Zhang, H., Zhi, L., Sun, S., Cui, S., & Yan, C. (2024). Phase-transition materials derived photonic metamaterials for passively dynamic solar thermal and coldness harvesting. *Heliyon, 10*(2), e23986. DOI: 10.1016/j.heliyon.2024.e23986 PMID: 38293359

Wu, L., Li, Y., Fu, Z., & Su, B. L. (2020). Hierarchically structured porous materials: Synthesis strategies and applications in energy storage. *National Science Review, 7*(11), 1667–1701. DOI: 10.1093/nsr/nwaa183 PMID: 34691502

Wu, Y., Wang, X., Yao, L., Chang, S., & Wang, X. (2023). Thermal Insulation Mechanism, Preparation, and Modification of Nanocellulose Aerogels: A Review. *Molecules (Basel, Switzerland), 28*(15), 5836. DOI: 10.3390/molecules28155836 PMID: 37570806

Xuan, L., Ju, Z., Skonieczna, M., Zhou, P. K., & Huang, R. (2023). Nanoparticles-induced potential toxicity on human health: Applications, toxicity mechanisms, and evaluation models. *MedComm, 4*(4), e327. DOI: 10.1002/mco2.327 PMID: 37457660

Xue, T., Zhu, C., Yu, D., Zhang, X., Lai, F., Zhang, L., Zhang, C., Fan, W., & Liu, T. (2023). Fast and scalable production of crosslinked polyimide aerogel fibers for ultrathin thermoregulating clothes. *Nature Communications, 14*(1), 8378. DOI: 10.1038/s41467-023-43663-8 PMID: 38104160

Xu, H., Li, X., Tong, Z., Zhang, B., & Ji, H. (2022). Thermal Radiation Shielding and Mechanical Strengthening of Mullite Fiber/SiC Nanowire Aerogels Using In Situ Synthesized SiC Nanowires. *Materials (Basel), 15*(10), 3522. DOI: 10.3390/ma15103522 PMID: 35629551

Yamasaki, S., Sakuma, W., Yasui, H., Daicho, K., Saito, T., Fujisawa, S., Isogai, A., & Kanamori, K. (2019). NanocelluloseXerogelsWith High Porosities and Large Specific Surface Areas. *Frontiers in Chemistry, 7*, 316. DOI: 10.3389/fchem.2019.00316 PMID: 31134187

Yang, D., Xu, P., Tian, C., Li, S., Xing, T., Li, Z., Wang, X., & Dai, P. (2023). Biomass-Derived Flexible Carbon Architectures as Self-Supporting Electrodes for Energy Storage. *Molecules (Basel, Switzerland), 28*(17), 6377. DOI: 10.3390/molecules28176377 PMID: 37687208

Yang, X., Jiang, P., Xiao, R., Fu, R., Liu, Y., Ji, C., Song, Q., Miao, C., Yu, H., Gu, J., Wang, Y., & Sai, H. (2022). Robust Silica-Agarose Composite Aerogels with Interpenetrating Network Structure by In Situ Sol-Gel Process. *Gels (Basel, Switzerland)*, *8*(5), 303. DOI: 10.3390/gels8050303 PMID: 35621601

Yang, Z., Wu, C., Shi, H., Luo, X., Sun, H., Wang, Q., & Zhang, D. (2022). Advances in Barrier Membranes for Guided Bone Regeneration Techniques. *Frontiers in Bioengineering and Biotechnology*, *10*, 921576. DOI: 10.3389/fbioe.2022.921576 PMID: 35814003

Yang, Z., Zhou, S., Feng, X., Wang, N., Ola, O., & Zhu, Y. (2023). Recent Progress in Multifunctional Graphene-Based Nanocomposites for Photocatalysis and Electrocatalysis Application. *Nanomaterials (Basel, Switzerland)*, *13*(13), 2028. DOI: 10.3390/nano13132028 PMID: 37446544

Yan, X., Wu, Y., Li, R., Shi, C., Moro, R., Ma, Y., & Ma, L. (2019). High-Performance UV-Assisted NO2 Sensor Based on Chemical Vapor Deposition Graphene at Room Temperature. *ACS Omega*, *4*(10), 14179–14187. DOI: 10.1021/acsomega.9b00935 PMID: 31508539

Yu, H., Li, H., Sun, X., & Pan, L. (2023). Biomimetic Flexible Sensors and Their Applications in Human Health Detection. *Biomimetics (Basel, Switzerland)*, *8*(3), 293. DOI: 10.3390/biomimetics8030293 PMID: 37504181

Yunus, G., Singh, R., Raveendran, S., & Kuddus, M. (2023). Electrochemical biosensors in healthcare services: Bibliometric analysis and recent developments. *PeerJ*, *11*, e15566. DOI: 10.7717/peerj.15566 PMID: 37397018

Zamparas, M., Tzivras, D., Dracopoulos, V., & Ioannides, T. (2020). Application of Sorbents for Oil Spill Cleanup Focusing on Natural-Based Modified Materials: A Review. *Molecules (Basel, Switzerland)*, *25*(19), 4522. DOI: 10.3390/molecules25194522 PMID: 33023166

Zhang, H., Zhang, C., Ji, W., Wang, X., Li, Y., & Tao, W. (2018). Experimental Characterization of the Thermal Conductivity and Microstructure of Opacifier-Fiber-Aerogel Composite. *Molecules (Basel, Switzerland)*, *23*(9), 2198. DOI: 10.3390/molecules23092198 PMID: 30200271

Zhang, H., & Zhang, Y. (2023). Rational Design of Flexible Mechanical Force Sensors for Healthcare and Diagnosis. *Materials (Basel), 17*(1), 123. DOI: 10.3390/ma17010123 PMID: 38203977

Zhang, L., Lei, Y., He, P., Wu, H., Guo, L., & Wei, G. (2022). Carbon Material-Based Aerogels for Gas Adsorption: Fabrication, Structure Design, Functional Tailoring, and Applications. *Nanomaterials (Basel, Switzerland), 12*(18), 3172. DOI: 10.3390/nano12183172 PMID: 36144967

Zhang, S., Yang, Z., Huang, X., Wang, J., Xiao, Y., He, J., Feng, J., Xiong, S., & Li, Z. (2022). Hydrophobic Cellulose Acetate Aerogels for Thermal Insulation. *Gels (Basel, Switzerland), 8*(10), 671. DOI: 10.3390/gels8100671 PMID: 36286172

Zhang, Y., Jiang, S., Xu, D., Li, Z., Guo, J., Li, Z., & Cheng, G. (2023). Application of Nanocellulose-Based Aerogels in Bone Tissue Engineering: Current Trends and Outlooks. *Polymers, 15*(10), 2323. DOI: 10.3390/polym15102323 PMID: 37242898

Zhang, Y., Poon, K., Masonsong, G. S. P., Ramaswamy, Y., & Singh, G. (2023). Sustainable Nanomaterials for Biomedical Applications. *Pharmaceutics, 15*(3), 922. DOI: 10.3390/pharmaceutics15030922 PMID: 36986783

Zhan, W., Chen, L., Kong, Q., Li, L., Chen, M., Jiang, J., Li, W., Shi, F., & Xu, Z. (2023). The Synthesis and Polymer-Reinforced Mechanical Properties of SiO2 Aerogels: A Review. *Molecules (Basel, Switzerland), 28*(14), 5534. DOI: 10.3390/molecules28145534 PMID: 37513406

Zhao, J., Yuan, X., Wu, X., Liu, L., Guo, H., Xu, K., Zhang, L., & Du, G. (2023). Preparation of Nanocellulose-Based Aerogel and Its Research Progress in Wastewater Treatment. *Molecules (Basel, Switzerland), 28*(8), 3541. DOI: 10.3390/molecules28083541 PMID: 37110772

Zheng, L., Zhang, S., Ying, Z., Liu, J., Zhou, Y., & Chen, F. (2020). Engineering of Aerogel-Based Biomaterials for Biomedical Applications. *International Journal of Nanomedicine, 15*, 2363–2378. DOI: 10.2147/IJN.S238005 PMID: 32308388

Zhou, Y. (2021). Artificial neural network-based smart aerogel glazing in low-energy buildings: A state-of-the-art review. *iScience*, *24*(12), 103420. DOI: 10.1016/j.isci.2021.103420 PMID: 34849473

Zhou, Y., Liu, Y., Zhang, M., Feng, Z., Yu, D. G., & Wang, K. (2022). ElectrospunNanofiber Membranes for Air Filtration: A Review. *Nanomaterials (Basel, Switzerland)*, *12*(7), 1077. DOI: 10.3390/nano12071077 PMID: 35407195

About the Contributors

Abel Inobeme is a senior lecturer and Head of the Department of Chemistry at Edo State University Uzairue, Nigeria. He holds a Nigerian Certificate in Education from the Federal College of Education Zaria, Nigeria; a B.Sc (Ed) Chemistry from Ahmadu Bello University Zaria, Nigeria; a Masters degree and PhD both in Analytical Chemistry from the Federal University of Technology Minna, Nigeria, and a Certificate in PhET Interactive Simulations from the University of Colorado Boulder. His research areas include environmental toxicological assessments, pharmacology, food analysis, wastewater treatment, and nanotechnology. He has 170 published works in top scientific journals to his credit. He has contributed more than 100 chapters in 60 edited books amongst top scientific publishers (Elsevier, Taylor and Francis, Springer Nature, Wiley, Sage, Royal Society of Chemistry, American Chemical Society amongst others). He is a reviewer and editor of some reputable scientific journals. He is the publication secretary of the Journal of Material Science and Engineering and also the current national publication secretary of the Material Science and Technology Society of Nigeria (MSN). He is also a member of the Chemical Society Nigeria (CSN), and the Nigerian Institute of Management (NIM). He was recently ranked among the top 500 scientists in Nigeria based on SCOPUS/SCIVAL output for the year 2023.

***

Mathew Adekoya is a condensed matter and nuclear science physicist with research interests in nuclear radiation, materials synthesis and characterization for device applications, and semiconductor devices.

Marian Osunde Agbugui has a PhD, MSc and B.Sc. in Biological Science from Ahmadu Bello University Zaria, Nigeria. She has numerous publications to her credit. She lectures in the Department of Biological Science at Edo State University Uzairue, Nigeria.

Esther Bernard is a lecturer in the Department of Chemical Engineering at Nassarawa State University Keffi, Nigeria. She holds a Ph.D. M.Eng and B.Eng in Chemical Engineering from the Federal University of Technology Minna, Nigeria

Kingsley Enerijiofi holds a Ph.D in Environmental and Public Health Microbiology from the University of Benin, Nigeria. He is an Associate Professor of Environmental / Public Health Microbiology and Nanotechnology and the current Ag. Dean, College of Basic and Applied Sciences, Glorious Vision University Ogwa Edo State Nigeria. His research interests is in Pollution control, Public Health, Nanotechnology and Environmental Sustainability. He teaches Environmental and Biotechnological based courses and has supervised a good number of degree research projects. He has published research papers in leading peer review local and international journals and also book chapters that are indexed in SCOPUS and Web of Science. He has organized international scientific conferences, has also attended and presented research papers. He is a member of council, Glorious Vision University Ogwa, Edo State Nigeria.

Goddidit Enoze holds B.Sc. Plant Biology and Biotechnology, MSc Phytomedicine and PhD in Phytomedicine all from the University of Benin, Nigeria. She has numerous publications to her credit.

Chinenye Eziukwu is a lecturer at Edo State University Uzairue, Nigeria. She specializes in Environmental Chemistry and Pollution Control. She has a Masters and B.Sc degree in Chemistry from the University of Ibadan

Jonathan Inobeme holds a B.Sc (Ed) degree in Geography with a second class upper division from Ahmadu Bello University Zaria, Nigeria. He also has a Masters and PhD from the same University. He has numerous publications to his credit. Jonathan Inobeme is a member of numerous professional bodies.

Tawakalitu Kelani is currently pursuing her PhD in Organic Chemistry at the Obafemi Awolowo University Ife. She has a Masters and B.Sc Degree from thesame university. She is a member of some professional bodies in her field. She has publication in local and international journals.

Maliki Muniratu has a Ph.D in Environmental and Analytical chemistry in 2017 from University of Benin, Benin City. She is presently Professor at Edo State University Uzairue, Edo State, Nigeria. Her research interest include: agricultural waste treatment and management, water analysis, nanotechnology, bioga and green technology. Her researches includes the synthesis and characterization of Chlorinated rubber seed oil (fire retardants) alkydes and Characterization of avocado and Africa pear oil

Index

A

B

C

E

F

H

I

L

N

P

R

S

T

www.ingramcontent.com/pod-product-compliance
Lightning Source LLC
LaVergne TN
LVHW081336300926
849161LV00008B/1988